Neural Network Model: Applications and Implications

Series Editor

Theodore Wasserman
Wasserman and Drucker PA
Boca Raton, FL, USA

Theodore Wasserman • Lori Drucker Wasserman

Disorders of Attention

Springer

Theodore Wasserman
Wasserman and Drucker PA
Boca Raton, FL, USA

Lori Drucker Wasserman
Wasserman and Drucker PA
Boca Raton, FL, USA

ISSN 2946-5753 ISSN 2946-5761 (electronic)
Neural Network Model: Applications and Implications
ISBN 978-3-031-78731-7 ISBN 978-3-031-78732-4 (eBook)
https://doi.org/10.1007/978-3-031-78732-4

This Springer imprint is published by the registered company Springer Nature Switzerland AG
The registered company address is: Gewerbestrasse 11, 6330 Cham, Switzerland

Preface

This book represents the fourth volume of our neural network series and the sixth book in a body of work that details both the challenges of the current behavioral nosology and the promise of neural network-based modeling for understanding disorders of behavioral and mental health. We began the series with the premise that there is both value and a need for understanding that there is a neural network system that underpins all human behavior, and that mind and body are neither separate, nor dissociable when it comes to understanding human behavioral (dys)function. Our initial work highlighted the challenges of the existing nosology exemplified in the DSM-5. We went on to explain a neural network basis to describe these dysfunctions and explain their neural network basis. In researching the constructs and neuroimaging, what we have come to is that these behaviorally defined constructs do not map efficiently onto neural imaging of network functioning. We are faced with accepting hypothetical constructs, and ignoring the science, or revising working hypothetical constructs to better match what neuroimaging can now better elucidate functionally. While they retain some clinical utility, there is need for greater understanding of the nature of human cognitive and emotional dysfunction.

We also recognize that the titles in the series project the idea that we can discreetly examine various aspects of functions as separate entities that are completely dissociable from other entities. That for example, Obsessive Compulsive Disorder is completely different from Attention Deficit Disorder. A network understanding of human cognitive and emotional functioning quickly disabuses us of this idea. We have learned that there are network commonalities among many disorders. Foundational networks, such as attention, the subject of this book, are recruited in every complex human behavior. We can talk about things like attention and motivation in the abstract, as independent constructs, but in reality, they never occur independently from one another when it comes to actual human behavior. We wrote the books this way so that we might be able to introduce the concepts of an integrated neural network system in a way that was comprehensible. After all, how many people might engage with a work entitled "Neural networks are integrated, and existing behaviorally defined constructs are not really dissociable?"

We would summarize by stating that the purpose of development is integration. That holds true for the development of neural networks. Beginning from largely discrete neural network hubs, human networks are developed by consistently linking these foundational centers with each other into increasingly complex, interconnected systems designed to solve the complex tasks that a typical person encounters in their daily life. These networks increasingly become interdependent and in many instances are created in a task-dependent manner. In relation to a specific task, they can't be pulled apart and assessed independently. To do so is antithetical to their creation.

Throughout these volumes we have attempted to square the existing behaviorally defined constructs of emotional and behavioral dysregulation with what is becoming known about human neural network creation, organization and functioning. It has become increasingly difficult to do so. One of the major reasons for this issue is the overly siloed nature of the research focus in neuropsychology. Much of what we research is constrained by researching the neuropsychological properties of the existing behavioral groupings. This leads to a process rich in self-affirmation but, perhaps, limited in research potential as it pertains to developing new and more effective intervention strategies. Being siloed means that we are not as exposed as we should be to the rich body of research on the same issues that exist in cognitive neuroscience, developmental neuroscience, genetics, and epigenetics. There is little doubt that incorporating ideas from these disciplines will challenge some of our most basic ideas regarding assessment and treatment of disorders having a network underpinning. We have identified some of these stress points in this volume.

It is not our intention to tear down the existing system but rather enrich it. We hope that giving the practicing neuropsychologist insights into the operation of the neural network system, we will enable them to develop new and effective techniques to help their clients. If, in the fullness of time, this process leads to a reimagining of what disordered attention, motivation, cognition, or emotional regulation should be called, we are open to it. Introspection, empirical investigation, and change are the hallmarks of science.

Boca Raton, FL, USA

Theodore Wasserman
Lori Drucker Wasserman

Contents

Chapter 1
Introduction

A mental health clinician wanting to ascertain what kind of diagnoses to give a person with an attentional problem would consult one of two resources: the Diagnostic and Statistical Manual 5 (DSM 5) or the International Classification of Diseases (ICD). These works represent the current clinical thinking on how attention is disordered in an individual. It might surprise you to learn that, apparently, the only designation for disordered attention is if you have some form of attention deficit hyperactivity disorder. The Diagnostic and Statistical Manual of Mental Disorders, 5th Edition (DSM-5) lists three main types of attention deficit hyperactivity disorder (ADHD):

Attention deficit hyperactivity disorder—Hyperactive/Impulsive Type
Attention deficit hyperactivity disorder—Inattentive Type
Attention deficit hyperactivity disorder—Combined Type
Attention deficit hyperactivity disorder—Other Specified
Unspecified attention deficit hyperactivity disorder

The ICD system recognizes the following:

Attention-deficit hyperactivity disorders
Attention-deficit hyperactivity disorder, predominantly inattentive type
Attention-deficit hyperactivity disorder, predominantly hyperactive type
Attention-deficit hyperactivity disorder, combined type
Attention-deficit hyperactivity disorder, other type
Attention-deficit hyperactivity disorder, unspecified type

In essence, these classifications suggest that any problem with f the complicated neural network system that subserves human attention is caused by some form of attention deficit hyperactivity disorder. For some time, this situation has seemed unrealistic.

T. Wasserman, L. D. Wasserman, *Disorders of Attention*, Neural Network Model: Applications and Implications, https://doi.org/10.1007/978-3-031-78732-4_1

In preparation for writing this book, we went to Google Scholar and typed the following search query: "disorders of attention." We certainly saw lots of research on attention deficit hyperactivity disorder, but we saw some other things as well. We saw papers on attention control, disorders caused by seizures and other neurologic conditions, and disorders associated with clinical disorders such as schizophrenia, anxiety, obsessive compulsive disorder, and autism. There was research on the things that cause disordered bottom-up or top-down attentional processing difficulties. There was also lots of research on the impact of the reward recognition system or emotional valence on attention. In other words, there were lots of ways attention can become dysfunctional. Certainly, there were many articles where disordered attention was referred to or defined as attention deficit hyperactivity disorder. In our view, this practice of lumping all forms of attentional problems under one, albeit comprehensive, label, confounded and continues to confound the research, making it much less likely that neural network-based subgroups and their functional properties, would be identified. Without such identification, research could not go forward, and potentially new and more effective treatments that addressed these subtypes would not get developed. It is just these issues that this volume set out to address. We hoped to demonstrate the complexity of the human attentional system and the myriad types of problems it can encounter. While we do not propose a new nosology, we do offer suggestions and examples of how such a neural network nosology might be created. Some beginning examples are provided.

In a recent workshop, we presented the idea that what we currently call primitive reflexes are in fact primary, or what we are terming "foundational reflexes." These innate processes are present at birth and are continuously integrated in ever more complex human cognitive and emotional operations. Although the focus of that presentation was on the motor-sensory system, the idea holds for the attention-sensory system as well. We encourage people to think of attention as a foundational process as opposed to describing these processes present at birth as "primitive." That is, the initial alerting and development of attention is foundational. We discussed how these elements become an integrated process in almost everything we do and discussed the idea that an early disruption or dysfunction in attentional processes could be a crucial component in the development of most, if not all, disorders of mental health as well as learning disabilities. We hope to demonstrate here that attention plays a critical role in most disorders of mental health and is etiologic for some, if not most, of them. We hope we have succeeded.

We also sought to demonstrate that attention doesn't exist in a vacuum. It requires a supporting cast. This is a point to which we will return many times; you cannot study attention without knowing the context. That is, just because you pay attention in one situation does not mean you will pay attention in another. It is not an independent entity. Nor is it likely a distinct entity, but rather it exists on a continuum, and it exists in relation to and in concert with other network-based processes. People always attend to something. Therefore, we can only measure attention in the context of what is being attended to. This is an idea with significant clinical implications. After all, can we really say a child's attention is disordered because he can't pay attention to tasks or lectures in school and comes home and misses many parental directives, but that same child comes home and plays on an Xbox for two hours straight with no breaks?

We have written extensively about the fact that we are at an inflection point in the evolution of our field when it comes to attention. While we have developed into a field with a wide fundamental acknowledgment of brain-based behavior, we continue to diagnose issues on the basis of behavior alone. We then perform cognitive gymnastics and explain the neural networks that underlie these behaviors. We will demonstrate that many of these explanations do not map onto actual network functioning at all. Therefore, our process is backwardin more ways than one. A new explanation of how things translate from network functioning to behavior is required.

Instead of just looking at patterns of behavior, we have begun the arduous journey of trying to understand how the human brain processes information and directs/controls behavior. Historically, in our attempt to understand it, we have made a significant error. We have attempted to measure behavior and the systems that subserve it one construct at a time. That is, constructs such as executive control and working memory. We treat them as if they are dissociable from one another, as if we can look at and measure them separately, that is, as discrete operational processes. We understand and appreciate the efforts that went into the ICD and how progressive an undertaking it was in its time. It was designed to categorize diseases. And if we are discussing a clear virus-borne or bacterially based disorder, one in which the resulting disorder is the clear result of a single entity, this system makes sense. However, human development does not follow that template. We would point out that the purpose of development, including from a neural network perspective, is integration of systems and function. Overtime, networks become interdependent as they cope with increasingly complicated task demands. This is especially true of attention, which, as we will see, is incorporated into almost everything a human does. As we develop, human attention, percentage-wise, becomes increasingly top-down in nature. That is, it is increasingly controlled by selection biases, reward value, and preferences, amongst many other things. As these other factors become increasingly important, it is impossible to argue that problems in those functions will not impact attention. Yet, as it stands today, the field does not recognize attention disorders that are based on functions such as aberrant designation of reward value or misallocated attentional resources. What would change if we do? For clarity, how would reframing the diagnostic characteristics of the aforementioned child playing Xbox over attending to his homework assignments alter both the diagnostic impression and treatment implications of these factors? We will see that these are real issues and real problems, and that the cognitive neuroscience literature outlines these types of attentional problems quite accurately.

Constructs exist based upon their definition, and as we shall see, regularly a significant portion of the available research conceptualizes issues, events, and constructs differently from what is known about the operation and integration of these processes, processes based upon neural networks which are recruited, or not, based upon the task or activity at hand. Reliance on said behavioral constructs has resulted in definitional confusion that abounds in many, if not all, psychological constructs that characterize mental health. We have stated our belief that this confusion clouds the effectiveness of research into many of these processes. The emphasis should be placed on processes.

We undertook this book with the desire to see whether the construct of attention would fall prey to the same problems. We have already written about the definitional problems associated with the diagnosis of attention deficit hyperactivity disorder (Wasserman & Wasserman, 2015) and suspected that similar issues would be inherent in the construct of attention as well. If this were a murder mystery, we would wait until the end to tell you what we found. This isn't a murder mystery, so we will tell you now. Problems with definitional inconsistency and lack of clarity abound in our psychological understanding of what is and is not attention or the cognitive processes that constitute its components. As a result, research into attention has been limited by these definitional constraints, and our understanding of how the brain works has been delayed. In addition, numerous forms of attention have been described. Table 1.1 summarizes some but not all of them.

It should be clear from the above that there are many processes that are involved in attention, and each of them can be disrupted in their own right. These processes are combined in complex human tasks, further complicating the picture. Attention can be disrupted in many ways. Mirsky and Duncan (2006) compiled a nosology which contained 16 disorders with disrupted attention, and the disrupted attention was not ADHD. These included Fragile X, sleep disorders including narcolepsy, breathing disorders, fetal alcoholism, lack of intellectual stimulation, lead intoxication, head injury, phenylketonuria, absence seizures, and malnutrition, among other difficulties. This serves to highlight the fact that the process of paying attention, the

Table 1.1 A taxonomy of attention (Oberauer, 2019)

	Controlled *Limited resource for controlling attention*		Automatic *Not resource limited*	
	Things/events	Goals/actions	Things/events	Goals/actions
Perceptual	Selective attention to locations, visual objects, events, and features. *Limited capacity for attending several channels*	Selective attention to ongoing actions, monitoring of action outcomes	Capture of attention by salient stimuli, stimuli learned to be relevant, or stimuli in the focus of attention of WM	Capture of attention by errors or unexpected difficulties
Non-perceptual	Attention to items in WM. *Limited capacity for maintenance ("storage")*	Attention to intended actions: selection of task sets, response selection. *Central processing capacity*	Involuntary retrieval from long-term memory; intrusive thoughts	Involuntary retrieval of task sets associated with the current stimulus; involuntary selection of response (e.g., Stroop, flanker task)

cortical subcortical connectivity, can be disrupted in many ways. A list of disorders that have attentional problems associated with them include the following:

- Asperger's Syndrome
- Attachment Disorders/Reactive Attachment Disorder (RAD)
- Autism Spectrum Disorder
- Biochemical Imbalances
- Borderline Personality Disorder
- Disruptive Mood Dysregulation Disorder (DMDD)
- Dyslexia
- Executive Function Difficulties
- Fetal Alcohol Syndrome
- Food Intolerances and Sensitivities
- Fragile X Syndrome
- Giftedness
- Hearing Problems—Including Auditory Processing Disorder
- Medical Conditions, e.g., Thyroid Disorder
- Mood Disorders (e.g., Anxiety, Depression, Bipolar Disorder)
- Non-verbal Learning Disabilities (NVLD)
- Obsessive Compulsive Disorder (OCD)
- PANDAS
- Post-traumatic Stress Disorder (PTSD)
- Schizophrenia
- Seizure Disorders
- Sensory Processing Disorder
- Sleep Disorders
- Specific Learning Difficulties
- Speech and Language Disabilities
- Substance Abuse
- Traumatic Brain Injury
- Vision Problems

The DSM 5 notes that "If these conditions are diagnosed and treated it may be that the symptoms associated with ADHD may disappear as a result." This statement recognizes that the diagnosis of ADHD is based on behavioral symptomology alone. While most of us are quite aware of this fact, we often forget about that when we talk about ADHD as having a neurological underpinning. We agree that there is a subset of individuals who have attention deficit hyperactivity disorder based on a discrete set of difficulties in the human neural network system that subserve attention. We ask you to consider the possibility that there is another subset of individuals who meet the behaviorally defined criteria for AHHD who do not. For example, there is another group of individuals who have difficulty assigning attention appropriately. Their attention is not broken; rather, it is misappropriated, misdirected, or misapplied. We will see that this interpretation of attention dysfunction accurately portrays the attentional difficulties associated with many disorders of mental health.

So, there are many different types of attention, and those types of attention can be disrupted in many ways. The fact that attention has numerous hypothetical types is variably defined, and that these multiple definitions obscure the implications of experimental work is not a new idea. Hommel et al. (2019) argued that no one knows what attention actually is and that its overuse and popularity in cognitive research have been detrimental to progress in understanding the processes involved. In fact, they state that "the concept of "attention" is one of the most misleading and misused terms in the cognitive sciences." They maintain the position that the term "attention" should be abandoned and the nature of the research in this area be reconceptualized to focus on the subsets of processes and mechanisms that lead to task-specific performance. This is because compartmentalizing "attention" and then searching for the "attentional system" hinders the development of a comprehensive understanding of human behavior because it ignores integrated, parallel, and reciprocal relationships among sensory, cognitive, and action processes.

Anderson (2011) argues that the slow progress in understanding attention can be ascribed to two general sources and one very specific one. The first fault is psychology's tendency to bifurcate or dichotomize constructs into being one thing or its opposite. We create arbitrary dichotomies and then research them as if they are facts. In the case of attention, the specific error is that we plurally define attention which leads to inconsistent usage and confusion. More importantly, attention has been reified; it has become a concrete concept that can act in a causal fashion. Anderson argues that attention never causes anything, because there is no such thing as attention, but there are many activities that are or can be accurately labeled attentional. To clarify, returning that child with the Xbox, his attention is not a solitary construct, nor is it always functional or not.

The third impediment to faster progress in understanding attentional effects is the equating of data with theory. For example, "studies reporting the patterns of functional brain activations or neuronal firing while subjects perform tasks that yield attentional effects will ultimately tell us much about the *how* of attention, but not the *what is* attention; that kind of understanding will require the combination of neuroscientific data with psychological theory."

Anderson concludes that the construct of attention was developed to describe a behaviorally observable set of processes. It was co-opted as a technical term to characterize these patterns of behavioral observations. These behaviors may or may not correspond to discrete neural processes. We have postulated that many constructs that we take for granted in modern neuropsychology can best be explained by conceptualizing them as a collection of task-specific processes engaged in by an integrated recruited network involved in problem solving. It is likely that attention is like that.

While we will talk about cognitive neuroscience, we are also practicing neuropsychologists. That means that we are applying the state of the art to assessing and understanding the behaviors being presented either in a child's daily living or under test conditions. As part of that job, we frequently assess the possibility that an individual, usually a child, does or does not have attention deficit hyperactivity disorder. To accomplish that task, we utilize the standard nosologies and appropriate psychometrics. While we recognize that the existing approach has merit and utility, we have, based upon our own work, developed some concerns. If there is a collection of neural network supported cognitive processes that collectively constitute what we now call attention, then it is likely that each of these processes can be disrupted, resulting in each instance in a failure of the individual to "pay attention." If that is true, then the question arises as to whether the resulting disruption is the same in every case. To put it another way, is all failure to attend due to attention deficit hyperactivity disorder? Is attention deficit hyperactivity disorder a specific dysfunction, or is it a description of what happens when one or more of the cognitive processes that are responsible for orienting a person to specific environmental stimuli fail? Or, given the developmental nature of integration, is it a weakness or failure somewhere in the neural network system? Are the behavioral manifestations the same regardless of the network being disrupted?

We will attempt to answer these and other questions. We will propose that, instead of looking at the behaviors that indicate a lack of attention, neuropsychologists, in specific, and clinicians, in general, should be looking at the various cognitive processes responsible for a person's ability to select an option for engagement. We will describe these processes and the neural networks that support them. We will note that despite overall "attention" being disrupted in every case, the specific network being disrupted leads to specific and unique attentional deficiencies. There is abundant research to support the idea that, when dealing with various clinical conditions, one's ability to engage in attentional processes should not be relegated to secondary consideration. We need to better elucidate how that person's attention is being allocated in order to better explain that person's current functioning and, if need be, what intervention might be applied. The implications of these proposed changes can be quite significant.

For example, we proposed that Attention Deficit Disorder represents an inefficiency of an integrated system designed to allocate working memory to designated tasks rather than the absence or dysfunction of a particular form of attention (Wasserman & Wasserman, 2015). We suggested that the failure to allocate sufficient working memory represented poor engagement of the reward recognition network with discrete circuits of learning and performance that control instrumental conditioning (learning). Efficient attention requires the interaction of these circuits. We concluded that for a significant percentage of individuals who present with Attention Deficit Disorder, their problems represent the engagement, or lack thereof, of the motivational and reward circuit as opposed to problems, or disorders of attention traditionally defined as problems with orienting, focusing, and sustaining. In other words, they had the ability to pay "attention" but were not sufficiently motivated to do so. This would lead to a set of behaviors that were different from those that characterized individuals, for example, with impulse control/go/no-go issues. This speaks to a need for a completely different understanding of "*how*" this child functions and a completely different explanation owed the parent about how to label or understand their child. Frankly, this is the role of the neuropsychologist—the explanation of brain-based behavior.

Think about attention deficit hyperactivity disorder, inattentive type. It can accurately be conjectured that people with this disorder can pay attention, but they choose to focus on something other than what they are required to focus on (motivation and reward network involvement) due to a lack of arousal from the presented material. This is quite different from the individual whose "attention" shifts rapidly with the presentation of novel stimuli in the environment, or hyperfocuses on novel stimuli and has a poor ability to inhibit the impulse to attend to the new stimuli. We will see that there is an argument to be made for ADHD inattentive type as a separate and distinct diagnosis from the other two presentations of the disorder.

We will look at the various core networks of the "attention" process and attempt to understand how a disruption of the smooth functioning of one of these networks leads to attentional deficiency. This will lead us to consider other types of attentional deficiencies that have, heretofore, been overlooked or lumped together into the mélange that is represented by the current understanding of attention deficit hyperactivity disorder.

References

Anderson, B. (2011). There is no such thing as attention. *Frontiers in Psychology, 2*, 246.

Hommel, B., Chapman, C. S., Cisek, P., Neyedli, H. F., Song, J. H., & Welsh, T. N. (2019). No one knows what attention is. *Attention, Perception, & Psychophysics, 81*, 2288–2303.

Mirsky, A., & Duncan, C. (2006). A nosology of disorders of attention. *Annals of the New York Academy of Science*. https://doi.org/10.1111/j.1749-6632.2001.tb05771.x

Oberauer, K. (2019). Working memory and attention – A conceptual analysis and review. *Journal of Cognition, 2*(*1*), *36*. https://doi.org/10.5334/joc.58

Wasserman, T., & Wasserman, L. (2015). The misnomer of attention deficit hyperactivity disoder. *Applied Neuropsychology: Child*, 1–7. https://doi.org/10.1080/21622965.2015.1005487

Chapter 2
Models of Attention

We have likely all, at one time or another, heard someone use the phrase "Pay attention." Or recognized that we "missed" something because we were attending to something else. Are these examples of "attention" the same? Is it the same process?

Let's start by stating that what we posit about the construct of attention is what the current authors, and others, have said regarding the lack of definitional uniformity in the understanding of many of the operations of human cognitive activities (Wasserman & Wasserman, 2013, 2015, 2017). That is, there is no uniformly agreed upon definition of attention, the effect of which impacts no less than clinical practice, neuropsychological assessment, and models that guide research on the subject.

What is common among most theories is the idea that attention is a "set of processes" guiding the enabling and selection of incoming perceptual information (Roda & Thomas, 2006). Just what these processes are, the number of processes involved, and the specific description of the neuropsychological operation of these processes vary from model to model. A review of some models will allow us to see how far along we have come in trying to identify this essential process. It also highlights the ongoing difficulties we face by lumping the construct and processes under one term. We will also see how the advent of neuroimaging furthers but also clarifies the picture. Finally, it should be noted that most of these earlier models constituted behavioral descriptions of observed processes as interpreted by the authors. There was no attempt, or ability, to map these constructs based on observation onto neural network structures or functions. More recently, this has changed, and we will learn that several of these early constructs do not map onto imaging of actual brain operation.

In 1991, Mirsky and colleagues (1991), in attempting to relate neuroanatomy and attention, utilized animal and human lesion studies and produced a seminal piece on attention, which came to be known as "The Mirsky Model." Their model was the result of consideration of impairment of attention as a characteristic of multiple disorders ranging from ADHD to schizophrenia. Mirsky et al. proposed three

T. Wasserman, L. D. Wasserman, *Disorders of Attention*, Neural Network Model: Applications and Implications, https://doi.org/10.1007/978-3-031-78732-4_2

elements—focus, sustain, and shift. He later expanded the model to include encode, focus/execute, and stabilize (Mirsky et al. 1999). While there are many updated analyses of attention, most models of attention attempt to better understand the process by analyzing segments of what in adults becomes a continuous, often parallel process. Sometimes models utilize arbitrarily created dichotomies to describe cognitive activities associated with attention (top down versus bottom up, divided versus focused attention, passive versus active attention, pre-attentive versus attentive) (Anderson, 2011). For example, Chun et al. (2011) distinguished attention according to types of information. In their model, external attention refers to the selection and modulation of sensory information (perceptual attention), whereas internal attention includes cognitive control as well as the contents of working memory, long-term memory, task sets, or response selection (central/reflective attention). Cognitive control mechanisms prioritize, independent of sensory modality, which perceptual information to select, encode, and maintain in working memory while suppressing distraction. While these processes are conceptually discernable, they are functionally heavily integrated and not readily dissociable. It is nevertheless important to understand the neural networks that subserve them so we can identify specific areas of dysfunction when they arise. In response, Anderson (2011) stated that "we need a new technical vocabulary that allows for attentional effects to be continuously distributed, rather than merely present or absent, and that provides a basis for quantitative behavioral predictions that map onto neural substrates." We agree.

There is a general consensus that attention as a construct is based on the idea that efficient/optimal interaction with the world around us requires that we attend to sources of information that help us form accurate beliefs about what is going on around us. Also, within the current consensus is that attention may represent two states. One in which we react and one in which we act. Obtaining optimum information can then be conceptualized as resulting from two very different sets of processes. The first set of processes concern themselves with our ability to select from multiple sensory sources of information those sensory streams that convey the most reliable information about states of the world and ascribe greater valence or weight to them. This has been called covert sampling. The second set of processes has a directed component wherein one actively manipulates the sensory system to overtly select the data we sense (Parr & Friston, 2019). This has been called overt sampling. The main difference between the process models is the idea that in the overt models, the receiver of information plays an active and, to some extent, predetermined role in selecting the information to be attended to. This key concept in understanding the difference between the covert and overt sampling of the sensory information we process results in models with a predictive (generative) component. This means that the brain possesses an internal model that tries to account for how sensations are generated by the external world (Friston, 2008).

Under this view, perception is a process of optimizing beliefs about the causes of sensations, while actions select new sensory data to test hypotheses about these causes. Research has suggested that there are neural pathways that correspond to the processes required by generative models. For example, Ungerleider and Haxby (1994) have suggested that connections from the visual cortex to the dorsal "where" and ventral "what" streams can be conceptualized as evidence in favor of spatial configurations or stimulus identities, respectively, the where and what (Ungerleider & Haxby, 1994). Those connections from frontal areas to posterior cortices may signal predictions about the "what" and "where" information that would be obtained as a consequence of action. Alternatively, they may modulate the gain of signals in these pathways. These two classes of models are equally valid, reflect distinct computational processes, and are mediated by different neurobiological structures (Parr & Friston, 2017). In other words, both processes might be operational within the same person. The current authors must then raise the question: Are these two processes discernable enough to warrant another term rather than referring to both a sensory information reactive process and an internally based predictive system simply as attention?

Nevertheless, as we continue to attempt to identify the workings of the process we call attention, newer models of attention and its component processes abound, and they might provide clues as to which of these components or processes are likely to be thought of as universal. In addition, current developments in neuroimaging are allowing us to better refine our understanding of brain behavior relationships within the context of functionally identifiable large-scale networks.

Types of Attention

Before we get to the various models of attention, it is beneficial to define terms relating to attention. Some of the models we will discuss deal with specific types of attention. Unfortunately, there is no universal consensus as the types of attention. The following list of terms is not exhaustive but will cover the points addressed by most of the models.

Sohlberg and Mateer (1987) divide attention into four and later, five components.

Arousal: a state of physiological activation or cortical responsiveness, associated with sensory stimulation and activation of fibers from the reticular activating system.

Sustained: Attention is the ability to focus on one specific task for a continuous amount of time without being distracted.

Selective attention: is the ability to select from many factors or stimuli and to focus on only the one that you want while filtering out other distractions.

Focus: Focused attention is the brain's ability to concentrate its attention on a specific target stimulus to the exclusion of others.

Alternating: The ability to switch focus between two different targets.

As mentioned above, Mirsky (1987), using factor analysis of neuropsychological tests, suggested four different types of attention and was one of the first to associate functioning with specific brain regions.

Focus-execute: associated with the inferior parietal lobes, basal ganglia, and superior temporal lobes.

Encode: reflects the ability to recall and reorder information stored in memory. Brain regions associated with encoding include hippocampus and amygdala.
Sustain: associated with activity in the Reticular Activating System tectum and the reticular thalamus.
Shift: found in the frontal lobes and the anterior cingulate gyrus.

Presaging a network understanding, Mirsky believed that even though the four factors appeared to be separate, in fact, all of these brain areas are linked together to form the human attentional system. Mirsky provided a model that served as the theoretical taxonomy for the neuropsychological battery of assessment of attention that is still in use.

Behavioral descriptions of certain "larger" types of cognitive processes include Visual Attention and Auditory Attention. In the research, they have been associated with specific brain regions. We view these activities as recruited networks that are developed to solve specific tasks.

Models of Attention

There are many ways to conceptualize the complex, heavily networked process we call attention. Most, if not all, models break the process up into behaviorally defined operationalized patterns of behavior. It is not at all clear that these "components" of attention actually exist discretely in function, but these descriptions of the components serve the utilitarian purpose of clarifying what actions need to occur in order for a person to orient themselves to a specific stimulus.

Earlier theories about attention were cognized without the benefit of today's neuroimaging. These earlier models relied upon lab studies to try to tease apart what the process of attending looked like. Today's advances allow us to attempt to associate behavior, structure, and function. To be clear, each of these components has neural networks that subserve them. These networks would be, in many instances, largely transitory, cobbled together to solve the specific task at hand. While it is important to understand this, it is also important that none of these highly task-specific networks would be standalone candidates for understanding things like attention deficit hyperactivity disorder. What would be helpful is to understand the functional and network commonalities between these models.

Selective Attention

There are three models that are associated with selective attention. These are the models of attention by Broadbent (1958), Treisman & Gelade (1980), and Deutsch & Deutsch (1963). They are also referred to as "bottleneck models of attention" because they explain how we cannot attend to all sensory input at one time in the conscious level.

Bottleneck models are all based on the idea that individuals have a limited number of attentional resources that are available for use at one time. Due to this limitation, environmentally available information and stimuli are processed in stages, or "filtered," resulting in only the most salient and important information being selected for perception and attention. An easy way to understand the ideas behind filtering theory and its importance to understand how we attend is to think about what happens when we go for dinner with friends in a crowded restaurant. Many restaurants these days seem to want to create a vibe of excitement and activity, resulting in a very noisy restaurant that requires the listener to parse out the background noise so as to be able to listen to the conversation of their dinner companion (Broadbent, 1958). How we do that—select what is important from all that noise—is the basis of filter theory.

There are, of course, many factors that determine what is selected and what is filtered out, and we will review these in separate chapters. What is important to understand here is that these processes determining the relative importance of various stimuli are crucial to the understanding of how attention works and where problems with attention might arise. What is also crucial is to consider that a single variable may not be the answer.

Broadbent Filter Model

Broadbent's filter model of attention (Broadbent, 1958) proposes a two-tiered system of processing. He proposed that the initial processing impacts all stimuli wherein the individual attends (extracts) to basic physical properties such as the pitch of the message and color of the stimulus. This information is stored in temporary memory. Because of the natural limitations to address cognitive loads, semantic features, those based on the meaning of the stimulus, require a "filter" to determine which stimuli are selected for further processing. The fact that certain information was restricted and filtered out led to the term "bottleneck." Initially, there were two essential questions that were asked: The first was what the differences were between the messages that were selected for attention and the messages that were selected to be filtered out. This might be conceptualized as what are the properties of the messages themselves that cause them to be selected.

The second question was whether the set of information that was ignored had some specific set of properties that caused that to happen. Initial research focused on the properties of the attended messages. Things like the loudness of the voice of a speaker were identified. The initial model posited that the information not attended to was lost to the listener and made no impact on them. This was guided by top-down influences as well as by bottom-up influences.

This position turned out to be problematic as people were, in fact, able to attend and remember some unimportant or filtered information. As we shall see, they might have been looking in the wrong place. The same message may have different levels of interest in the listener depending on the property of the listener, their history, and their values. To be sure, there are messages that have a high value of their own (a tornado alert or hurricane warning), but most messages are attended to based on their value to the listener.

Broadbent proposed two stages in this attention process. The first stage was based upon the clearly discrete properties of the message; the second concerned itself with the actual "information" in the message. It was this second stage that was limited by its processing capacity and had to be filtered for the salient parts. This second stage protected the receiver of information from being "overloaded" by the amount of information coming at them. He did posit that when little information is required for processing, one can attend to and process multiple stimuli simultaneously.

This leads directly to one of the characteristics of certain children with attention deficit hyperactivity disorder. They appear to have filtering problems in that background and noise are equally attractive to them. Take, for example, "little David," who is watching the butterfly outside instead of apparently listening to the math teacher yet is able to answer the question "David, what did I just say?" when called upon. We could posit that this type of disorder is distinct from other attentional deficiencies and deserves its own ICD code: Attention filtering disorder.

Attenuation Model of Treisman

The attenuation model of Treisman (Treisman, 1969) keeps both the idea of an early selection process and the apparatus by which physical cues are used as the primary point of inequity. The attenuation model posits that there is a decrease in the perceived loudness of an unattended message. This decrease in volume results in the message not being loud enough to reach its threshold, unless it has a very low threshold to activate it to begin with (like calling your name), or there is a general momentary decrease for all messages. So, in contrast with Broadbent's filter model, the filter now "attenuates" the unattended information instead of filtering it out totally. The model focused on a supposed structural limitation on stimulus processing, out of which attentional selectivity emerges. In a major contribution, Treisman introduced the concept of a threshold to explain how some words came to be heard in the unattended channel with greater frequency than others. In other words, some of the background noise made it through and was processed less completely but nevertheless processed.

Mesulam's Model

Mesulam (1981) considered directed attention to be both specialized and redundant. It was composed of four domains of attentional regulation: a reticular component that regulated both arousal and vigilance, the posterior parietal lobe that provided a sensory map of the world, the cingulate cortex that regulated motivational behavior, and the frontal cortex that regulated motor action programs designed to explore the surrounding environment.

The Model of Posner and Dehaene

Posner and Dehaene (1994), early proponents of brain networks, posit that attention is characterized by three major functions: orienting to sensory stimuli, executive functions, and maintaining the alert state. These network descriptions were based on localization indicated in imaging studies of the behavior manifestations of the operation of each of these networks.

The executive control network operates to control goal-directed behavior, target selection, error detection, conflict resolution, and inhibition of automatic responses. This executive control network seems to include the midline frontal areas, including the anterior cingulate gyrus, supplementary motor area (SMA), and portions of the basal ganglia. Of note is that the basal ganglia are believed to facilitate communication between the executive control system and areas associated with other attentional networks.

The second network, called the alerting network, is responsible for establishing a state of readiness to react and maintaining vigilance over that state. It is composed of Brodmann's area, the right frontal lobe, especially the superior region, the right parietal lobe, and the locus coeruleus. The third network is responsible for orienting attention. The recruited brain regions include the parietal lobe and the fusiform gyrus.

Presaging our discussion, Posner and Dehaene note that attention deficit hyperactivity disorder has been identified as the result of disruptions in all of the conceptualized networks. Swanson et al. (1998) characterized ADHD as a combination of executive attention and alerting dysfunction. Barkley (1998) attributed ADHD exclusively to deficits in executive functions, defined as problems with the internalization of self-directed speech and the control of emotional arousal and motivation. This kind of attribution becomes easier to understand when it is recalled that ADHD is defined as a cluster of behaviors taken from a longer list of qualifying behaviors. Different behavioral deficit sets from within that broader range of qualifying deficits can lead to the same diagnosis.

The Role of Large-Scale Brain Networks

A good way to understand the distributed nature of attention function is to look at the role of the various large-scale brain networks in its operation. Joyce and Hrin (2015) provide a concise summary of the various large-scale brain networks and their role in attention. Based on the work of Yeo et al. (2011), they identify seven functional brain networks that have involvement in the allocation of attentional resources. Interestingly, these seven areas incorporate about 97% of the cortical area of the brain. The networks identified were as follows.

Frontoparietal Network

This network is a cognitive control/working memory network that is activated when a stored rule or information based on past experiences is needed to guide the selection of a target goal for action. In turn, this large-scale network recruits subnetworks including the dorsolateral PFC and the anterior cingulate cortex, ACC, the inferior parietal lobe, and numerous subcortical structures in order to accomplish its task. The system plays a large role in determining goal-directed behavior.

The Dorsal and Ventral Networks

These networks play a central role in attentional processing. The dorsal network is involved in the management of spatial attention and in the shifting of attention. In other words, this is the system that identifies where objects in the environment are located and then brings forward information on how to use them. An example of this would be looking for a screwdriver in a tool drawer.

The ventral network is responsible for actual object identification and also identifies how to use it. This network determines salience, that is, the "what" that makes this particular item notable, one of the key determinants of attentional allocation. This network also interrupts ongoing behavior when something more salient enters the stimulus field. Anatomically, the temporo-parietal junction, the supramarginal gyrus, the frontal operculum, and the anterior insula comprise this network. These two networks, the dorsal and ventral networks, often work in tandem with each other.

The Visual Network

This network receives input from both the visual cortex and, despite its name, auditory sensory-processing centers. It has important connections to the superior parietal lobe and the intraparietal sulus, which are in turn connected to the dorsal attention network, outlined above. This network is important for sustained attention and suppressing attention to irrelevant stimuli.

The Motor Network

The motor network is the first functional neural network. It is composed of the primary, supplementary, and pre-motor cortex in addition to the sensory cortex, putamen, thalamus, and cerebellum. This is the foundational network, and in the model we are proposing, it is the network that eventually ties all the other networks together into a functional and, in many ways, non-dissociable, whole.

The Default Mode Network

This network is active when external stimuli are at a minimum. It projects to the dorsomedial cortex when a person is engaged in internal cognitive processes and the medial temporal lobe when a person is trying to project themselves into the future. Collectively, this line of modeling projects itself onto the vertical brain modeling that forms the foundation model for this book. The network is composed of the anterior medial PFC, posterior cingulate, dorsomedial prefrontal, and medial temporal areas.

The Limbic Network

We will talk about the limbic system in detail in a separate chapter. This system underlies reward recognition and motivation. It is composed of the dorsal ACC and the bilateral insulae, which provide for signaling for salient events, including errors.

Historically, the division of attention into its component parts was based on behavioral descriptions of apparent attentional functions. These descriptions, at times, may have correlated with actual brain organization, but for the most part, they were descriptive of activities and not how the brain was actually facilitating attentional processing. The network model represents the first attempt at understanding how the brain is organized to facilitate attention that represents actual brain function. Understanding how these networks interact with one another is critical to understanding attentional processing.

Biased Competition Theory (Kim & Kastner, 2019)

Biased competition theory concerns itself largely with the process of selective attention in the visual field. It is used both in computational modeling of attention and in modeling human neural network behavior. The model posits that each object in the visual field competes for cortical representation, and cognitive processing is based on both top-down and bottom-up information processing. This theory hypothesizes that the process of visual processing can be biased by other mental processes, such as bottom-up and top-down systems, both of which prioritize certain features of an object or whole items for attention and further processing. This competition can be biased, often toward the object that is currently attended in the visual field, or alternatively toward the object most relevant to behavior. We will discuss this model in some detail, and it contextualizes the role of reward recognition within a framework of other biasing factors such as size, shape, or relevance.

Biased competition is used to systematically study the neural basis of visuospatial attention. Visuospatial attention refers to selective attention allocated at a specific location in visual space. It involves filtering out the visual input from unattended locations, which is unwanted and potentially distracting. There are three basic principles of the model: that representation in the visual system is competitive; that both top-down and bottom-up biasing mechanisms influence the ongoing competition; and that competition is integrated across brain systems. There is considerable research evidence that support all three principles (Beck & Krastner, Top-down and bottom-up mechanisms in biasing competition in the human brain, 2009).

The first and most fundamental prediction of biased competition theory is that objects compete for neural representation in the visual cortex. These multiple stimuli are present at the same time within a neuron's receptive field (RF). They are not processed independently but interact with each other in a mutually suppressive way, essentially vying for supremacy and attention. These suppressive interactions among multiple stimuli present simultaneously in the visual field indicate these stimuli are competing for representation by single neurons in the visual cortex. The winner is determined by the presence of biasing factors, both top-down and bottom-up, that determine which stimuli get attended to and which do not. The following example might be instructive. I wake up one morning to discover my coffee maker is broken, and I decide to buy a new one. I go to the appliance store, and there is a wide array of coffee makers available. They vary in size, color, shape, price, and complexity of function, to name just a few variables. My preferences for each of these variables represent biases that narrow my visual scanning of the field. Novelty might be another bias; perhaps I see a maker in a different shape or with different functions that "catch my eye." By comparing the images against my biases, I continually narrow my search. A critical prediction for the neural implementation of biased competition theory is that the competitive interactions should be strongest when the stimuli activate the same local region of cortex (Beck & Kastner, Stimulus context modulates competition in human extrastriate cortex, 2005). If stimuli are competing for representation by a particular neuron, then the competitive interactions should be most apparent when the stimuli fall within the RF of that same neuron.

The second tenet of biased competition theory is that the competition can be controlled by introducing *biases* that favor the processing of a particular stimulus at the expense of competing stimuli. Moreover, biased competition theory posits multiple biasing mechanisms. For us, this is important because each of these biasing mechanisms has the potential to capture and regulate attention, and a defect in the process or an idiosyncratic biasing mechanism has the potential to distort attentional regulation.

The third tenet of biased competition theory is the hypothesis that selection of a target object emerges through the integration of many separate competitive systems, such that when an object gains dominance in one system (e.g., visual cortex), it will tend to gain similar dominance in other systems that are interfaced with the visual cortex for the task at hand (Everling et al., 2006). We highlight this here to support the idea that multiple networks contribute to task resolution.

There are models of attention that address specific areas of attention, some of them emanating from computational modeling. For example, Feature Integration Theory (Treisman & Gelade, 1980) posits that in visual search paradigms, the feature search and the conjunction search are conducted in parallel and sequentially, respectively. This idea of parallel processing as opposed to serial processing is an important element for understanding human attention processing in general. Feature search is a visual search process that focuses on identifying a previously requested target among distractors that differ from the target by a unique visual feature such as color, shape, orientation, or size (such as seen on symbol search takes on psychological tests). A conjunction search task is a process of identifying a desired object when it is in close proximity to objects that share some of its same features. For example, finding a red crayon in a box of crayons containing differing shades. Humans engage in conjunction search tasks on multiple occasions every day.

The theory has two major hypotheses. The first is that each single feature (e.g., color, edge orientation, size, motion, and orientation) in the visual field is processed individually by the corresponding specific module, spatially in parallel. This results in humans not being required to pay attention to a specific location to search for targets. Secondly, in the case of conjunction search, humans have to localize and integrate the results from multiple modules of a unique feature. This integration requires an attentional shift to a specific location. Because of this attentional shift, the conjunction search is processed sequentially.

The Model of Guided Search (Wolfe et al., 1989)

This is the model that introduced top-down knowledge on characteristics of target stimuli in visual search. The guided search model uses an activation map computed from both bottom-up activation based on the existing feature-search process and top-down activation based on the characteristics of target stimuli. Using feature search bottom-up activation is guided by saliency considerations in each feature channel (e.g., color, orientation) in input stimuli.

The top-down activation is obtained based on the correlation between input and target stimuli with regard to each of the features. For instance, if the target stimulus has features consisting of "black" and "vertical line," the black regions in the top-down color map and the vertical line regions in the top-down orientation map are activated. The activation map is achieved by summing up the top-down and bottom-up activation maps. Accordingly, the focus of attention is oriented in the order of activation levels. Guided search is a model that explicitly implements the characteristics of target stimuli in visual search and serves as a basis for more recent top-down computation models (Kimura et al., 2013).

Contextualizing Models of Attention

To a greater or lesser degree, all of the theories we have reviewed have in common that they are based on behavioral observations of individuals who are in the process of attending. Focusing, orienting, salience, and attending all constitute behavioral descriptions of observed behavior. They are described as segmented and dissociable activities that can be assessed individually and in isolation from other aspects of attentional processing. Further, extant research has attempted to identify specific brain areas and networks that control these disparate processes. The question is whether or not these descriptions and assumptions are correct. Do they characterize correctly the underlying, white matter network connectome of the human brain? As we will describe in the chapter on the neural network properties of attention in the human brain, these behaviorally based descriptions do not map when connectome studies are done. Furthermore, the various components of attentional processing are clearly not dissociable from each other in the adult human brain. In other chapters, we will discover that they are highly integrated and interdependent upon each other. We are not the first to suggest this idea.

There is research to support a highly integrated attentional system (Stuss, 2006). In summarizing his findings, he concludes "Our lesion studies focusing on attention have demonstrated impairments in distinct processes due to pathology in different frontal regions, implying fractionation of the supervisory system." However, when task demands are manipulated, it becomes evident that the frontal lobes are not just a series of independent processes. Increased complexity of task demands elicits greater involvement of frontal regions along a fixed network related to a general activation process. For some task demands, one or more anatomically distinct frontal processes may be recruited. In other conditions, there is a bottom-up nonfrontal/frontal network, with impairment noted maximally for the lesser task demands in the nonfrontal automatic processing regions, and then as task demands change, increased involvement of different frontal (more "strategic") regions, until it appears all frontal regions are involved. With other measures, the network is top-down, with impairment in the measure first noted in the frontal region and then, with changing task demands, involving a posterior region. Adaptability is not just a property of the frontal lobe; it is the fluid recruitment of different processes anywhere in the brain as required by the current task" (p. 261). This is not to say that there are not specific and identifiable subroutines because there are (Tsuchida & Fellows, 2013). These routines are recruited in a task-dependent manner (Mohr et al., 2020) and participate in numerous attentional paradigms. The overall conclusion is that focusing on one task may involve many but not all of the subroutines utilized when focusing on another type of task. The same can be said of the other aspects of attention described in the behavioral models.

Looking at the evolution of the human brain might provide us with an understanding of why this might be so. The human brain evolved from its more primitive form to its current structure by incorporating primitive structures and adding to and expanding them over its evolutionary course (Koziol & Budding, 2009; Hofman, 2014). This development was based on adaptation and supported the integration of later evolving higher-order cognitive activities (such as value-based motivation) with primitive sensory and motor functions. It follows from this that the functions of those primary/foundational areas would be retained in every area that became integrated with them over the course of evolution. These foundational elements, such as the primitive visual and motor systems, would therefore likely serve as the core elements of processes such as attention. As the requirements for attention increased with the growing complexity of the tasks that human beings are confronted with, these primitive/primary elements would have remained at their core. Thus, the historic models of attention and its organization would not represent the neural network operation of the human brain, and newer models would be required to correctly represent human attentional neural network functioning.

Ultimately, two things are true. First, the position of the current authors that attention is best understood as the processes involved in the operation of an integrated system (Wasserman & Wasserman, 2015) is being verified on a daily basis. The position was that the integrated system includes reward saliency, motivation, working memory, and other factors falling under the heading of executive function. The position also held for an emotional component as an equal variable.

The second truth is that neuroimaging has allowed for the understanding that this process is a highly integrated and finely tuned interplay of neural networks which respond to both external and internal variables.

The current authors understand and respect the fact that in order to understand the concept of attention, the current state of investigation requires some deconstruction of the construct. The current authors also wish, however, to stress the clinical and investigatory need to better describe the construct itself in order to better provide explanations of a person's functioning and to better apply appropriate interventions.

References

Anderson, B. (2011). There is no such thing as attention. *Frontiers in Psychology, 2*, 246. https://doi.org/10.3389/fpsyg.2011.00246

Beck, D., & Kastner, S. (2005). Stimulus context modulates competition in human extrastriate cortex. *Natuure Neuroscince, 8*(8), 1110–1116. https://doi.org/10.1038/nn1501

Beck, D., & Krastner, S. (2009). Top-down and bottom-up mechanisms in biasing competition in the human brain. *Vision Research, 49*(10), 1154–1165. https://doi.org/10.1016/j.visres.2008.07.012

Broadbent, D. (1958). *Perception and communication*. Oxford University Press.

Chun, M. M., Golomb, J. D., & Turk-Browne, N. B. (2011). A taxonomy of external and internal attention. *Annual Review of Psychology, 62*(1), 73–101.

Deutsch, J. A., & Deutsch, D. (1963). Attention: Some theoretical considerations. *Psychological Review, 70*(1), 80.

Everling, S., Tinsley, C., Gaffan, D., & Duncan, J. (2006). Selective representation of task-relevant objects and locations in the monkey prefrontal cortex. *European Journal of Neuroscience, 23*, 2197–2214. https://doi.org/10.1111/j.1460-9568.2006.04736.x

Friston, K. (2008). Hierarchical models in the brain. *PLoS Computational Biology, 4*, e1000211.

Hofman, M. (2014). Evolution of the human brain: When bigger is better. *Frontiers in Neuroanatomy, 8*, 26. https://doi.org/10.3389/fnana.2014.00015

Kim, N., & Kastner, S. (2019). A biased competition theory for the developmental cognitive neuroscience of visuo-spatial attention. *Current Opinion in Psychology, 29*, 219–228. https://doi.org/10.1016/j.copsyc.2019.03.017

Kimura, A., Yonetani, R., & Hirayami, T. (2013). Computational models of human visual attention and their implementations: A survey. *IEICE Transactions on Information and Systems, E96-D*(3), 562–578.

Koziol, L. F., & Budding, D. E. (2009). *Subcortical structures and cognition: Implications for neuropsychological assessment*. Springer.

Mirsky, A. F. (1987). Behavioral and psychophysiological markers of disordered attention. *Environmental Health Perspectives, 74*, 191–199.

Mirsky, A. F., Anthony, B. J., Duncan, C. C., Ahearn, M. B., & Kellam, S. G. (1991, June). Analysis of the elements of attention: A neuropsychological approach. *Neuropsychology Review, 2*, 109–145.

Mirsky, A. F., Pascualvaca, D. M., Duncan, C. C., & French, L. M. (1999). A model of attention and its relation to ADHD. *Mental Retardation and Developmental Disabilities Research Reviews, 5*(3), 169–176.

Mohr, K., Carr, N., Georgel, R., & Kelly, S. (2020). Modulation of the earliest component of the human VEP by spatial attention: An investigation of task demands. *Cerebral Cortex Communications, 1*(1), tgaa045. https://doi.org/10.1093/texcom/tgaa045

Parr, T., & Friston, K. J. (2017). Working memory, attention, and salience in active inference. *Scientific Reports, 7*(1), 14678.

Parr, T., & Friston, K. J. (2019). Attention or salience?. *Current Opinion in Psychology, 29*, 1–5.

Posner, M. I., & Dehaene, S. (1994). Attentional networks. *Trends in Neurosciences, 17*(2), 75–79.

Roda, C., & Thomas, J. (2006). Attention aware systems: Theories, applications, and research agenda. *Computers in Human Behavior, 22*, 557–587.

Sohlberg, M. M., & Mateer, C. A. (1987). Effectiveness of an attention-training program. *Journal of Clinical and Experimental Neuropsychology, 9*(2), 117–130.

Stuss, D. (2006). Frontal lobes and attention: Processes and networks, fractionation and integration. *Journal of the International Neuropsychological Society, 12*(2), 261–271. https://doi.org/10.1017/S1355617706060358

Treisman, A. (1969). Strategies and models of selected attention. *Psychological Review, 76*, 282–299.

Treisman, A., & Gelade, G. (1980). A feature-integration theory of attention. *Cognitive Psychology, 12*, 97–136.

Tsuchida, A., & Fellows, L. (2013). Are core component processes of executive function dissociable within the frontal lobes? Evidence from humans with focal prefrontal damage. *Cortex, 49*(7), 1790–1800. https://doi.org/10.1016/j.cortex.2012.10.014

Ungerleider, L., & Haxby, J. (1994). 'What' and 'where' in the human brain. *Current Opinion in Neurobiology, 4*, 157–165.

Wasserman, T., & Wasserman, L. D. (2013). Toward an integrated model of executive functioning in children. *Applied Neuropsychology: Child, 2*(2), 88–96.

Wasserman, T., & Wasserman, L. (2015). The misnomer of attention deficit hyperactivity disoder. *Applied Neuropsychology: Child, 4*, 1–7. https://doi.org/10.1080/21622965.2015.1005487

Wasserman, T., & Wasserman, L. D. (2017). *Neurocognitive learning therapy: Theory and practice*. Springer International Publishing.

Wolfe, J., Cave, K., & Franzel, S. (1989). Guided search: An alternative to the feature integration model for visual search. *Journal Experimental Psychology: Human Perception and Performance, 15*(3), 419–433.

Chapter 3
Do All Disorders of Attention Result in Attention Deficit Hyperactivity Disorder?

This is a book about what happens when one or more of the processes that subserve attention does not work as intended. As we shall see, there are many neural network-subserved processes that contribute to attention, and many ways these processes can be disrupted or, alternatively, fail to develop. As clinicians, we understand that when clinicians talk about disrupted attention, they most often refer to attention deficit hyperactivity disorder. Are these two ideas synonymous? Do we understand everything we need to know about how attention does not work properly by simply understanding the research behind ADHD?

There is little doubt that multiple dysfunctional cognitive processes are found in individuals with ADHD. These include abnormalities of neural responses underlying anticipation or planning of actions, target selection, filtering of distracting information, working memory, response selection and/or inhibition, novelty detection, error signaling, reward evaluation, and feedback-mediated decision-making processes, among others (Bush, 2010). Currently, disruptions of one or multiples of these networks that cause disrupted attention are grouped into a few behaviorally defined categories that have little direct reference to the network systems that underlie them (American Psychiatric Association, 2013). ADHD is a developmental syndrome whose cardinal signs are inattention, impulsivity, and hyperactivity. Currently, the disorder is classified by several types, including an inattentive form, a relatively rarer purely hyperactive form, and a combined type that is the most common form and features both inattention and hyperactivity. Inattention, or the inability to direct and maintain selective attention to motivationally relevant tasks, is an essential feature of the disorder. Impulsivity refers to acting rashly without apparently thinking of the consequences, such as blurting out answers in class. Hyperactivity refers specifically to excessive motor activity.

We do not take issue with the clinical relevance of these behaviorally defined disorders. It is also true that there is more than one way to reach a behavioral outcome. Understanding the network functioning of the contributing networks may in

T. Wasserman, L. D. Wasserman, *Disorders of Attention*, Neural Network Model: Applications and Implications, https://doi.org/10.1007/978-3-031-78732-4_3

fact elucidate differing subsets of these behaviorally defined disorders that may better inform research and lead to more effective understanding and research. A couple of examples may clarify this concept. Cognitive disorganization is routinely cited as one of the main indicators of inattention. Cognitive disorganization can include those processes relating to executive functions, such as impulse control and working memory, as well as non-executive functions, such as memory and reaction time. This cognitive disorganization can be produced in many different ways. Disorganization can result from poor general motivation or reflect an impaired ability to value long-range goals over short-term rewards or reflect problems with reward valuation. Motivational issues alone do not account for all of the ways that cognitive dysfunction can arise. Even when individuals are motivated, disorganization can manifest from an inability to plan ahead and order the required steps to achieve a goal. Disorganization could also result from poor working memory skills that impair the ability to maintain goals and/or plans in mind, or it could be produced by an inability to adjust behavior to meet changing contingencies (Bush, 2010). Clearly, these differing etiologies are important, and treating them may require differing forms of intervention. The focus on the behavioral level blurs potentially important distinctions.

Attention as Value Allocation

Similarly, we have proposed that a subset of attention-deficit disorder inattentive type represents an inefficiency of an integrated network system designed to allocate working memory to designated tasks rather than the absence or dysfunction of a particular form of attention. This inefficiency in the allocation of working memory represents, in part, poor engagement of the reward recognition network with neural networks subserving learning and performance that control instrumental conditioning (learning). For this subset of individuals who present on a behavioral level with attention-deficit disorder inattentive type, their problems represent the engagement, or lack thereof, of the motivational and reward recognition network as opposed to problems, or disorders of attention traditionally defined as problems with orienting, focusing, and sustaining. We demonstrated that there is an integrated system of working-memory allocation that responds by recruiting relevant aspects of both cortex and subcortex to the demands of the task being encountered. In this model, attention is viewed as a gating function determined by novelty, flight-or-fight response, and reward history/valence affecting motivation (Wasserman & Wasserman, 2015). Subsequent research has substantiated the role of the reward recognition network for understanding disrupted attention (Anderson, 2017). There are even examples of the networks that subserve this process. These networks support the idea that reward history reciprocally guides attention to reward stimuli. One of these networks is the complex and differentiated locus coeruleus norepinephrine (LC-NE) system that is involved with different behavioral and cognitive substrates of reward and attention. Specifically, the LC receives reward-related sensory, perceptual, and visceral inputs, releases NE, glutamate, dopamine, and various neuropeptides, forms reward memories, drives attentional bias, and selects behavioral strategies for reward. This research points to preclinical and clinical studies that have found that abnormalities in the LC-NE system are involved in a variety of psychiatric conditions marked by disturbed functions in reward and attention (Zhang et al., 2023). This line of research understands that conventional models of attention posit a theoretical dichotomy between goal-directed top-down attentional control and stimulus-driven bottom-up attentional control. However, this conventional theory leaves a gap in the understanding of attentional selection and allocation. There is, in fact, a third category termed "selection history" that indicates that the history of attentional deployments can bias attentional selections (Awh et al., 2012). Specifically, past rewarding experiences detached from current goals and physical salience can posit a lasting effect on attentional control (Anderson, 2021).

These are just two examples of the way attention can be disrupted. There are others. There can be little doubt that these examples represent two very different groups of people and that repair of these deficient processes might involve very different types of remedial activities. A clinical example might be instructive.

A Clinical Example of Attentional Dysregulation

Recently, a young lady, aged 12, was referred to be evaluated for the presence of attention deficit disorder that was potentially disrupting her ability to perform in school. She had just entered middle school, and completing work was increasingly difficult. She reported to her parents that she was having problems concentrating, and they wanted to address the problems expeditiously. A diagnostic interview was conducted during which the child freely discussed her problems with attention and concentration. Checklists were provided to the child, the parents, and selected teachers. All of the checklists came back with significant endorsements of behaviors consistent with a diagnosis of attention deficit hyperactivity disorder. It was assumed that the youngster was bright and that her symptoms had been masked until now. She attended an academically competitive school and had a history of good grades and no conduct problems. Everything pointed in the direction of ADHD, the inattentive type, and the parents seemed content with that understanding.

There were some issues, however, that were not consistent or at least raised areas of concern. The first set of concerns was related to the fact that there were also significant endorsements for social anxiety, general anxiety, and problems related to autism spectrum disorder. When considering a diagnosis of ADHD, these issues are not necessarily disconfirming. Children with ADHD frequently have problems relating to their peers, display significant anxiety, and can be socially anxious as a result. The second area of concern is that the parents had a teacher from the year before the current year complete a set of questionnaires. That teacher reported no difficulties and provided a rating score that was all in the average range. During the feedback session, the parents were asked about this disparity, and the mother responded that in the prior school year, the child's grandmother had died and that the child was very close with her grandmother. Upon reflection, the parents observed that they had noticed a change in the child's behavior after the death of her grandmother.

There is no question that the child's ability to attend was disrupted and that she had met clinical thresholds on checklists of ADHD. There is no doubt that she is having problems focusing and completing work. We suppose it might be argued that the child is having an emotional problem and does not have ADHD at all. But if ADHD is a behavioral diagnosis and she meets the behavioral criteria at the current time, why doesn't she have it? It can be argued that her attention networks are not disrupted and that the problems stem from the dysregulation of her emotions. If we consider the third way outlined above, it becomes easy to see that her attentional system is certainly dysregulated by factors such as reward valuation and anxiety. If we were to argue that this is not implied when the term ADHD is used and that ADHD is somehow just neurological, why do we not have types of ADHD that reference disruptions in top-down or bottom-up processing specifically? These are certainly "neurological" conditions.

To be clear, there are many ways attention gets disrupted. Either attention deficit disorder references all of them or it references some of them. If it references solely those of "neurological" origin, the classification system needs to have many more than the three categories that currently comprise the diagnosis. If an integrated network understanding of human brain functioning has any value, it is in understanding that these "emotional" things are not dissociable from "neurologic" things. There are neural networks that subserve both. They work together to produce sophisticated behavioral outcomes. The quicker our research understands this, the quicker we develop more sophisticated and efficacious solutions to the clinical problems that confront us in our everyday practice.

Historical Models of Disordered Attention

There have been models developed that posit that there are disorders of attention that are quite different and dissociable from attention deficit hyperactivity disorder. Many of these models presaged/anticipated/ confirmed network modeling but did not explicitly describe neural networks. Mirsky et al. (1999) presented a model of disordered attention that included ADHD but was not limited to it. The model was not based on a direct relationship with network modeling but was instead based on performance on neuropsychological tests. The model divided attention into a set of particular states/capacities, including focusing, sustaining attention, encoding, shifting attention, and executive functions. They also included the concept of stability of response. This model incorporated the idea that attention can be disrupted in a number of ways, and not all the ways it became disrupted were reflective of ADHD. In fact, their model indicated that a number of these core processes were involved in ADHD. In addition, they believed that ADHD may be due to "immature" brain development of any or all of the neurological systems that subserve the abovementioned capacities. Mirsky discussed his belief that many of these core capacities were coordinated or linked into a region based integrated system in the brain. This basic idea of networking was not elaborated and did not rely on neural networks to explain how that might be accomplished. Later analysis did layer the model onto what was known about the operation of large-scale neural networks (Koziol et al., 2014).

The model contemplated attentional deficiencies arising either from faulty development or direct damage to the subserving brain regions. In their model, individuals with ADHD were impaired in some manner in at least three and perhaps all four of the core elements. Problems with encoding, shifting, and sustaining attention were identified. Children with ADHD were seen as impulsive and poorly regulated. In other words, individuals with ADHD represented instances wherein most if not all of the components of the attentional system were dysfunctional. They were the extreme cases.

What of those instances where just one of the processes was dysfunctional? Could a person have trouble solely with shifting, for example, and what would the impact of that problem be on overall attentional capacity? Certainly, inability to shift, or perhaps the opposite, hyperfocus, is seen in a variety of disorders (obsessive-compulsive disorder, autism) where overall attention is also impaired. These disorders impact the application of attention as well. For example, OCD appears to be associated with an attentional bias favoring threatening information as well as reduced levels of cognitive inhibition (Muller & Roberts, 2005). These disorders demonstrate frontostriatal pathophysiology, differentiated by the fact that ADHD is characterized by frontostriatal hypoactivity and OCD with hyperactivity. In addition, OCD and ADHD appear to share similar neuropsychological impairments, especially in executive functions (Abramovitch et al., 2011). An argument can be made that these and other disorders are a reflection of different aspects of the same networks' dysfunction.

It is legitimate to ask whether, in reality, such a circumstance, wherein we could isolate one specific area of attention for inspection, could occur. Certainly, it would be possible to design tasks where shifting is a major contributing factor, but is it possible to design tests where it is the only factor being manipulated? Following on, were those tasks that were determined to assess one factor really only assessing that factor to the exclusion of all other elements of attention? Are these elements dissociable in real life? These are but a few of the questions that arise when considering a network-based model.

Other theories of attention postulate differing mechanisms and processes. Theories of selective attention posit that learning about a stimulus depends on attending to a specific stimulus. Attending is represented in 2-stage models that argue that people switch in analyzers as well as learning stimulus-response associations, represented in a formal model by the incorporation of a stimulus-specific learning-rate parameter that contains equations describing changes in the associative strength of stimuli as a result of stimulus and reward strength. Previous theories of selective attention have also assumed that (a) people learn to attend to and ignore relevant and irrelevant stimuli (i.e., that attention may increase or decrease depending on the correlation of a stimulus with reinforcement); and (b) there is an inverse relationship between the probabilities of attending to different stimuli. That is, an increase to one stimulus is accompanied by a decrease in attention to others (Mackintosh, 2016).

Attention can be classified into two distinct functions: bottom-up attention, referring to attentional guidance based upon purely externally driven factors to stimuli that are salient because of their inherent properties relative to the background; and top-down attention, which relies on internal guidance of attention based on prior knowledge, willful plans, and current goals. Over the past few years, insights on the neural circuits and mechanisms of bottom-up and top-down attention have been developed through neurophysiological experiments. Attention affects the mean neuronal firing rate as well as its variability and correlation across neurons. Although distinct processes mediate the guidance of attention based on bottom-up and top-down factors, a common neural apparatus, the frontoparietal network, has been found to be essential in both types of attentional processes (Katsuki & Constantinidis, 2014).

Attentional Functioning Occurs Along a Continuum

The overarching point here is that attention, however described, is a complex process that is subserved by a complex neural network system. Hypothetically, this process can be disrupted at any point in this system. It is hard to imagine that the resulting disruption would be the same in every case, and that means that there are more types of attentional deficiencies than just the three types of attentional issues detailed in the current clinical literature. There are also indications that symptoms vary in severity, and individuals with milder symptoms may have attentional problems but are not classified with ADHD. There is some research support for this idea (Lubke et al., 2009). Their data, based on factor analysis of maternal rating scales, indicated that an attention syndrome (as defined by the Child Behavior Checklist problems scale) varies along a severity continuum of mild to moderate to severe attention problems across three dimensions: hyperactivity, inattention, and nervousness. Their data further indicated that children diagnosed with ADHD are at the extreme of the continuum. This, of course, makes sense given the nature of rating scales, where the critical cutoff for clinical significance is often the top 2½%. The data provides evidence that ADHD symptoms exist on a continuum rather than as discrete yes/no categories. The logical extension of this finding is that there are many children who experience behavioral problems leading to inattention, who have some or all of the symptoms associated with ADHD, and who do not meet the clinical criteria for ADHD. The question arises as to what we should call this group and more to the point: is their attention disordered or not? Their mothers clearly think that it is. There are other ideas associated with the idea that ADHD symptoms occur on a continuum. The implications of this pattern of findings suggest the desirability to de-emphasize categorical conceptualizations of ADHD, produce evidence to better inform risk-benefit ratios of interventions along a spectrum of symptom and functional severity, and more coherently triage and arrange service delivery on the basis of symptom and functional severity rather than artificial diagnostic categorizations (McLennan, 2016).

The idea of a continuum of attentional functioning and the idea of a heavily integrated and non-dissociable neural network supporting attentional processes taken together implies that attentional disorders likely share many characteristics with other disorders. For example, there is a line of research (van der Meer et al., 2012; Johnson et al., 2014) that suggests that ADHD and Autism Spectrum Disorder (ASD) can be conceptualized not as distinct neurodevelopmental conditions but rather as representatives of one primary condition with a similar etiology. The hypothesis underlying this theory posits ASD to be a manifestation of the most "severe" subtype on one end of the continuum, while mild ADHD would be located on the other end of this hypothesized continuum. If this hypothesis holds, its theoretical implication would be that ASD cannot exist without ADHD. This hypothesis has been somewhat modified in that ADHD and ASD represent something in between, neither ends of a continuum nor completely separate disorders (Deserno et al., 2023). They share aspects of attentional deficiency and differ on others. We will see that there are specific attention deficits associated with ASD that are essential to the behavioral problems associated with the disorder. We will also see how it is possible to build a nosology based on the characteristics of the attentional dysfunction rather than a set of behavioral problems. Finally, on a clinical level, we will learn that it is very difficult to search for disorders of attention that are not ADHD. The extremely heterogeneous grouping of many types of attention problems under this one rubric compromises future research into the clinical implications of a network model of attention deficiency.

References

Abramovitch, A., Dar, R., Hermesh, H., & Schweiger, A. (2011). Comparative neuropsychology of adult obsessive-compulsive disorder and attention deficit/hyperactivity disorder: Implications for a novel executive overload model of OCD. *Journal of Neuropsychology*. https://doi.org/10.1111/j.1748-6653.2011.02021

American Psychiatric Association. (2013). *Diagnostic and statistical manual of mental disorders* (5th ed.). American Psychiatric Association.

Anderson, B. (2017, March). Reward processing in the value-driven attention network: Reward signals tracking cue identity and location. *Social Cognitive and Affective Neuroscience, 12*(3), 461–467. https://doi.org/10.1093/scan/nsw141

Anderson, B. A. (2021). An adaptive view of attentional control. *American Psychologist, 76*(9), 1410.

Awh, E., Belopolsky, A., & Theeuwes, J. (2012, August). Top-down versus bottom-up attentional control: A failed theoretical dichotomy. *Trends in Cognitive Sciences, 16*(8), 437–443. https://doi.org/10.1016/j.tics.2012.06.010

Bush, G. (2010). Attention-deficit/hyperactivity disorder and attention networks. *Neuropsychopharmacology, 35*, 278–300. https://doi.org/10.1038/npp.2009.120

Deserno, M., Bathelt, J., & Groenman, A. (2023). Probing the overarching continuum theory: Data-driven phenotypic clustering of children with ASD or ADHD. *European Child & Adolescent Psychiatry, 32*, 1909–1923. https://doi.org/10.1007/s00787-022-01986-9

Johnson, M., Gilga, T., & Charman, T. (2014). Annual research review: Infant development, autism, and ADHD – Early pathways to emerging disorders. *The Journal of Child Psychology and Psychiatry*. https://doi.org/10.1111/jcpp.12328

Katsuki, F., & Constantinidis, C. (2014). Bottom-up and top-down attention: Different processes and overlapping neural systems. *The Neuroscientist, 20*, 509–521. https://doi.org/10.1177/1073858413514136

Koziol, A., Joyce, A., & Wurglitz, G. (2014). The neuropsychology of attention: Revisiting the "Mirsky Model". *Applied Neuropsychology: Child, 3*(4). https://doi.org/10.1080/21622965.2013.870016

Lubke, G. H., Derkes, E., van Bijsterveldt, T., & Boomsma, D. (2009). Maternal ratings of attention problems in ADHD: Evidence for the existence of a continuum. *Jorunal of the American Academy of Child & Adolescent Psychiatry, 48*(11), 1085–1093. https://doi.org/10.1097/CHI.0b013e3181ba3dbb

Mackintosh, N. J. (2016). A theory of attention: Variations in the associability of stimuli with reinforcement. *Psychological Review, 82*(4), 276–298. https://doi.org/10.1037/h0076778

McLennan, J. (2016). Understanding attention deficit hyperactivity disorder as a continuum. *The Official Journal of the College of Family Physicians Canada, 62*(12), 979–982.

Mirsky, A. F., Pascualvaca, D. M., Duncan, C. C., & French, L. M. (1999). A model of attention and its relation to ADHD. *Mental Retardation and Developmental Disabilities Research Reviews, 5*(3), 169–176.

Muller, J., & Roberts, J. (2005). Memory and attention in obsessive–compulsive disorder: A review. *Journal of Anxiety Disorders, 19*(1), 1–26. https://doi.org/10.1016/j.janxdis.2003.12.001

van der Meer, J., Oerlemans, A. M., Van Steijn, D. J., Lappenschaar, M. G., De Sonneville, L. M., Buitelaar, J. K., & Rommelse, N. N. (2012). Are autism spectrum disorder and attention-deficit/hyperactivity disorder different manifestations of one overarching disorder? Cognitive and symptom evidence from a clinical and population-based sample. *Journal American Academy Child Adolescent Psychiatry, 51*(11), 1160–1172. https://doi.org/10.1016/j.jaac.2012.08.024

Wasserman, T., & Wasserman, L. (2015). The misnomer of attention deficit hyperactivity disoder. *Applied Neuropsychology: Child*, 1–7. https://doi.org/10.1080/21622965.2015.1005487

Zhang, Y. C., Xin, Y., Peng, B., & Liu, S. (2023). Norepinephrine system at the interface of attention and reward. *Progress in Neuro-Psychopharmacology and Biological Psychiatry, 125*, 110751. https://doi.org/10.1016/j.pnpbp.2023.110751

Chapter 4
Attention Deficit Hyperactivity Disorder: Genetics, Neural Networks, and Other Factors

This is a book about attention. It is more specifically a book about the various associated disorders that can occur when the complex neural network system that underpins attention does not work as it should. It speaks to the fact that there are many ways that the attentional system can be disordered. This is in some way at odds with current clinical practice. When we as clinicians speak about attention, we likely first consider the categorical diagnosis of attention deficit hyperactivity disorder (ADHD). We would be remiss if we did not address attention deficit hyperactivity disorder. After all, from a clinical perspective, ADHD is the iconic condition that most laypeople and clinicians alike think of when considering disordered attention.

It is the major thesis of this book:

- That there are many ways that attention may become disordered.
- There are many ways in which attention may present when disordered.
- And not all of them fundamentally represent the same construct we call attention.

First, it then should be clear that the current authors hold that not all disorders of attention should be assumed to represent the disorder we call attention deficit hyperactivity disorder. That is, the symptomology may meet criteria for said diagnosis, but is that both accurate and valid?

Second, what is this developmental disorder we are calling attention deficit hyperactivity disorder? The DSM-ll initially termed this disorder Hyperkinetic Reaction of Childhood. Beginning with DSM-lll, the term Attention Deficit Disorder with or without hyperactivity came into use. As of the DSM-IV, the terminology has been ADHD with inattentive presentation, ADHD with hyperactive presentation, or ADHD with combined presentation, representing both hyperactivity and inattentiveness. Based upon the current iteration, the DSM-5, these then are now coded as Attention Deficit Disorder Hyperactive Type, Attention Deficit Disorder Inattentive Type, and Combined Presentation. Does this current iteration then represent three disorders of attention or one disorder with multiple presentations? The DSM 5

T. Wasserman, L. D. Wasserman, *Disorders of Attention*, Neural Network Model: Applications and Implications, https://doi.org/10.1007/978-3-031-78732-4_4

(American Psychiatric Association, 2013) considers this one disorder with multiple diagnosable presentations, which can, however, overlap (combined type). The current authors are, as are others, concerned with the purported distinction, or lack thereof, between the umbrella disorder and the subtype presentations.

The current functional model has a couple of basic characteristics. It functionally lumps multiple symptoms under the diagnostic category descriptors, implying that most, if not all, problems with attention are subsumed into one heterogeneous category and that the construct of attention is universally understood as a discrete entity. While it is not our intention to innumerate all of the issues with the current DSM system, we would highlight a couple as it pertains to our discussion of attention. As noted, the DSM lists an array of symptoms associated with this construct. The lists are behavioral. The diagnosis is then made on the basis of behavioral criteria. The system is silent as to etiology and provides no information as to the network properties of the disorder. This system of diagnosis has its issues. Among them are:

A. Should we be basing this neurodevelopmental disorder on the presentation of behaviors which may or may not be present in every iteration of the disorder and may be confounding?
B. Should we be presenting this disorder as a discrete disorder when, as clinicians, we know that it is part of a "larger" issue? That is, attention is not a discrete domain. It presents, in potentially disordered ways, across psychopathologies.
C. In addition, how do we fail to address subthreshold symptoms?
D. Do these behavioral classifications map onto the known neural networks that subserve attention?

Part of the problems we face as clinicians is that insurance coverage, a major driver in the health system, requires a (1) diagnosis and (2) diagnosis indicative of sufficient magnitude to cause behavioral and emotional dysfunction. This encourages the current system, as one of the "positive" aspects of the current system is that it is easily grasped by both clinicians and clients and provides a reasonable shorthand for clinicians.

The current system also generates additional questions regarding both the diagnostic process and the underlying theory. Is there an alternative to this current process of picking and choosing from the large array of behaviors? Is there a way to conceptualize a greater understanding of the "why" of a person presenting with said behaviors, as well as a better descriptor of a particular subtype or presentation of the disorder? And think about how the better targeting of etiology might assist in better targeted interventions.

One such alternative speaks to the recognition of the possibility of a multiplicity of symptom patterns and potential causes for any particular disorder. This concept is referred to as Etiologic Heterogeneity (Girault et al., 2024). It addresses how different subtypes of a disorder are shaped by different risk factors, for example, a genetic variant, and how multiple methodologies and transdiagnostic approaches can lead to a better understanding of the underlying factors inherent in various clinical phenotypes. In fact, the current findings in research, noting genetic overlap among many diagnostic labels, are providing insight into both the neural and psychological bases of psychopathology and are subsequently at the core of the ongoing NIMH RDoC project (Clark, 2017). The current authors support the ongoing efforts of various researchers and policy makers to try to address the current shortcomings in our diagnostic system. More specifically, it is clear to the current authors that, for the purposes of research and treatment, devising a system which includes dimensionality, intensity of both primary and extended symptoms, and developmental and environmental considerations of symptom presentation is necessary.

The current writers understand that, as clinicians or neuropsychologists, it may not be feasible to consider all of the factors which should be looked at when determining diagnostic issues. This is still aspirational and may be limited to the halls of research. However, might this not also direct us toward a greater understanding of a neural network model of ADHD? And might not the overlay of a combination of empirically validated cognitive treatment combined with a functional understanding of neural network models, as they involve cognitive and affective domains, support a greater understanding of the totality and primacy of the presenting symptoms, making treatment more efficacious. How does this help the clinician and the researcher? Clearly, through this greater development of understanding of the neurobiological underpinnings of a disorder comes a greater ability to target more streamlined and efficient interventions. It can take the person from considering something they "have" to how they are impacted and how they can utilize interventions to address the primary and secondary issues.

We believe that a thorough understanding of the neural network properties of attention will support these contentions. We also believe that even a rudimentary understanding of a neural network model in a clinical setting allows for a greater, more sophisticated, and more practical understanding of ADHD. This, in turn, provides a clinician with a valuable foundation of information to help a parent, teacher, etc., truly better understand the behavioral manifestations of ADHD that brought them to seek help in the first place. More on this later.

For now, in order to expand on this idea, we will first review the current behaviorally defined diagnostic criteria for ADHD and its recognized subtypes. From there, we will review the problems with the current diagnostic nosology and go on to describe the network properties of ADHD in specific and attention in general.

From the outset, it is important to note, as stated above, that from a clinical perspective, ADHD is a behaviorally defined and identified (discrete) disorder based upon behavioral criteria. This is quite different from the extant research that suggests that even though the precise etiology of ADHD is not fully understood, "it is considered as a multifactorial and heterogeneous disorder with many contributing factors such as heritability, auxiliary to neurodevelopmental issues, neuroconnectivity issues, severe brain injuries, neuroinflammation, consanguineous marriages, premature birth, and exposure to environmental toxins" (Yadav et al., 2021). These various factors, along with other factors, can produce behaviorally similar but etiologically different disorders. This leads to the question: Does etiology matter? Does dimensionality matter? We believe that it does. We believe that the greater our discerning of the confluence of genetics, underlying network involvement, the role of environment, and developmental stages will all facilitate our ability to discern the presentation patterns we are seeing and be better able to provide the client/patient with an understanding beyond simply of what a patient "has." It will allow for better targeted intervention. We also believe that clarifying these patterns of presentation will facilitate cleaner and more meaningful research.

Current Diagnostic Criteria for Attention Deficit Hyperactivity Disorder

Clinicians use the guidelines in the American Psychiatric Association's Diagnostic and Statistical Manual, Fifth Edition (DSM-5), (American Psychiatric Association, 2013) to diagnose ADHD. These guidelines consist of behaviorally defined descriptors/symptoms. The authors of the DSM 5 have attempted to address some concerns with respect to heterogeneity and presentation. They have placed the disorder under neurodevelopmental disorders. However, they state that "no biological marker is diagnostic for ADHD," a statement with which the current authors concur. However, the DSM 5 also states that "meta-analysis of all neuroimaging studies do not show differences between individuals with ADHD and control subjects" (p. 73) and "do not reference biological markers or neuroimaging." This statement represents many of the difficulties of concern that cause us to suggest the necessity, in our opinion, of bringing our diagnostic models in line with neuroimaging and weighting studies, many of which are discussed throughout this volume. The problem here is that there are inconsistencies in the meta-analysis due to the very heterogeneity of the disorder. Consider the following: The inattentive subtype is prevalent in about 18.3% of the total number of persons diagnosed, while hyperactive/impulsive is prevalent in approximately 8.3% and combined represents 70% (Magnus, 2023). We must ask: What happens when 70% of the construct is represented by two constructs? What intervention should take precedent? And of equal importance, what happens to the statistical power of our research?

Taking a closer look, the current symptoms are grouped into two major categories: symptoms of (in)attention and symptoms of hyperactivity/impulsivity. In an effort to recognize the impact of developmentally based differences, the number of symptoms required for an ADHD diagnosis varies by age group as follows:

Six or more symptoms of inattention and/or hyperactivity-impulsivity for children up to age 16 years.

Five or more symptoms of inattention and/or hyperactivity-impulsivity for adolescents ages 17 years and older and adults.

The symptoms of inattention are as follows:

- Often fails to give close attention to details or makes careless mistakes in schoolwork, at work, or with other activities.
- Often has trouble holding attention on tasks or play activities.

- Often does not seem to listen when spoken to directly.
- Often does not follow through on instructions and fails to finish schoolwork, chores, or duties in the workplace (e.g., loses focus, sidetracked).
- Often has trouble organizing tasks and activities.
- Often avoids, dislikes, or is reluctant to do tasks that require mental effort over a long period of time (such as schoolwork or homework).
- Often loses things necessary for tasks and activities (e.g., school materials, pencils, books, tools, wallets, keys, paperwork, eyeglasses, mobile telephones).
- Is often easily distracted.
- Is often forgetful in daily activities.

The symptoms of impulsivity/hyperactivity are as follows:

- Often fidgets with or taps hands or feet, or squirms in seat.
- Often leaves seat in situations when remaining seated is expected.
- Often runs about or climbs in situations where it is not appropriate (adolescents or adults may be limited to feeling restless).
- Often unable to play or take part in leisure activities quietly.
- Is often "on the go" acting as if "driven by a motor."
- Often talk excessively.
- Often blurts out an answer before a question has been completed.
- Often have trouble waiting their turn.
- Often interrupts or intrudes on others (e.g., butts into conversations or games).

In addition to the above, the following criteria should also be met:

- Several inattentive or hyperactive-impulsive symptoms were present before age 12 years.
- Several symptoms are present in two or more settings (such as at home, school, or work; with friends or relatives; in other activities).
- There is clear evidence that the symptoms interfere with, or reduce the quality of, social, school, or work functioning.
- The symptoms are not better explained by another mental disorder (such as a mood disorder, anxiety disorder, dissociative disorder, or personality disorder). The symptoms do not happen only during the course of schizophrenia or another psychotic disorder.

We get our first indication of the trouble that this behavioral nosology causes in trying to elucidate ADHD, as it is likely a very heterogeneous population of individuals who receive the diagnosis. One of the clearest issues is that given the total number of symptoms that are available, it is quite possible that there would be numerous combinations of symptoms that lead to the same diagnosis. In fact, from the nine symptoms of inattention, you can make 60,840 unique combinations. The same number can be made from the nine hyperactive/impulsive factors. Hypothetically, many of these varying combinations of symptoms might represent differing disorders. At the very minimum, they may reflect differing forms of the same core disorder. In addition, we should consider a neurodevelopmental disorder from a developmental perspective; is the disorder subject to differences based upon the developmental age of the individual? Longitudinal studies suggest that developmentally based trajectories should be considered, with Posner et al. (2020) positing early onset (3–5 years), middle childhood (6–14 years) onset with persistent course, middle childhood onset with adolescent offset, and adolescent or adult onset (16 years and older). Their data suggested that the childhood ADHD group, as compared to the adult ADHD group, indicated no overlapping sets, with 90% of adult ADHD cases indicating no history of childhood ADHD. Of importance, the adult ADHD group did not show test-based neuropsychological deficits in childhood nor in adulthood.

In addition, many difficult-to-quantify associated signs of ADHD exist, and these symptoms are frequently also associated with other disorders. Let us take, for example, disorganization. Disorganization is often cited as one of the main sequelae of inattention. Disorganization may refer to functional disorganization such as "often does not follow through on instructions and fails to finish homework, chores or duties in the workplace (loses focus or is easily sidetracked)" or "has difficulty organizing tasks and activities (misses details)" or may refer to cognitive disorganization such as "difficulty sustaining attention in tasks or play activities" or "fails to give close attention to details or makes careless mistakes" (APA, 2013). As a refocus on these descriptors highlights, the categories of cognitive and functional blur. In addition, "disorganization" can result from many different causes. Disorganization can result from poor general motivation or can reflect an impaired ability to value long-range goals over short-term rewards, or it can result from other activities (i.e., playing video games) being more highly valued. Even when people are motivated, disorganization can result from an inability to plan ahead and order the necessary steps to accomplish a goal. Apparent disorganization could also result from poor working memory skills that impair the ability to maintain goals and/or plans in mind, or it could be produced by an inability to adjust behavior to meet changing contingencies (Bush, 2010). Disorganization should also be considered in light of affective disorders such as depression. Certainly, we can largely agree that a difficulty in attending to details, distractibility, and a general failure to follow through with tasks and chores are often inherent in depression.

Given the importance of neural network integration to the models discussed herein, we wish to also point out how the emphasis on diagnostic criteria for ADHD has altered. Specifically, the National Comorbidity Survey reported that adults with ADHD are three times more likely to develop major depressive disorder (MDD), six times more likely to develop dysthymia, and more than four times more likely to have any mood disorder (Kessler et al., 2006). In fact, the second edition of the DSM in 1968 identified the disorder as Hyperkinetic Reaction of Childhood, which places a clear emphasis on the "excessive" motor component (American Psychiatric Association, 1968).

This is not to say that the current diagnostic criteria for ADHD do not reliably diagnose a condition. They do. Considerable clinical experience and research evidence now support their use: these criteria have yielded reliable and consistent prevalence estimates across different national and cultural settings and have shown evidence of robust association with neurobiological measures (Rosales et al., 2015). However, the question we need to ask is, while the condition can be reliably diagnosed, is it a valid construct, and is there discriminative validity to the diagnostic process?

Reliability speaks to agreement. We all agree we are seeing some version of some of these, both behavioral or clinical symptoms and cognitive symptoms. However, given the breadth of symptoms from which to select, what construct are we each looking at?

Asked another way, while the condition can be reliably diagnosed, is it any longer a valid diagnostic construct? Specifically, there is also evidence that the current diagnostic criteria are all inclusive and represent, arguably, several different and perhaps distinct conditions.

We highlight that among the three identified forms of ADHD there are structural, genetic, and white matter differences that suggest that they might be better considered as separate disorders. So now we must consider the effects on research as noted above when the largest subtype subsumes the two more "distinct" subtypes in combination with the fact that current imaging suggests that indeed we are not looking at a single construct. Let's look more closely at that.

Structural Brain Changes and Associated Symptoms

Mu et al. (2022), recognizing the different cognitive and behavioral profiles as well as varying neurobiologies associated with the inattentive and combined types of ADHD, wanted to assess whether these subtypes represented separated brain structural changes. They looked at differences in the cortical thickness, area, and volume, and subcortical volume and hipposubfield volume in young people between the ages of 8 and 14 who fell into the typically developing class, the ADHD Combined Grouping or the ADHD Inattentive Type classification. They used T1-weighted MRI data and obtained structural magnetic resonance imaging scans; diagnostic and demographic information were obtained from 234 individuals. The sample contained 138 Typically Developing individuals, 37 of whom were ADHD-Combined, and 59 were ADHD-Inattentive patients. As mentioned above, they measured cortical thickness, area, and volume, subcortical volume, and hipposubfield volume. They found that ADHD-Inattentive patients showed milder clinical symptoms but more serious cognitive "injury" (Full scale IQ and Verbal scale IQ) than ADHD-Combined patients. Importantly, dissociable structural brain changes were found in different subtypes of ADHD, particularly in terms of decreased subcortical volume in ADHD-Combined patients compared with Typically Developing individuals. They concluded that clinical symptoms were mainly related to smaller caudal anterior cingulate thickness and left pallidum volume, whereas verbal IQ "injury" was strongly correlated with the finding of a smaller insula area. These findings are indicating significant differences in clinical symptoms and gray matter damage between ADHD-Combined and Inattentive patients, with supporting evidence of heterogeneity in the ADHD-Inattentive subtype as well as evidence of brain structure differences. Mu et al. highlight some additional functional symptom differences of importance. Specifically, relating the findings for the inattentive type group having lower overall IQ and also often being found to have reading and language deficits. This is important data as it relates strongly to directing intervention needs.

Although limited, there are other imaging studies of ADHD that have investigated the differences between subtypes of ADHD.

Semrud-Clikeman et al. (2017) also set out to compare the subtypes of ADHD. They expected that children with ADHD would present with smaller volumes of the anterior cingulate cortex and caudate. Volumetric imaging did indeed find bilaterally smaller volumes of the caudate and anterior cingulate cortex in children diagnosed with ADHD Combined Type compared with ADHD-Inattentive children. Also of note, parent ratings of attention and hyperactivity predicted right anterior cingulate cortex volume, whereas hyperactivity was associated with right caudate volume. They conclude that different structures are involved in ADHD subtypes and also suggest that this may reflect different phenotypes.

Diamond (2005) also looked at subtype differences. Diamond found that the frontal-parietal circuit is affected in ADHD-Inattentive patients, whereas the primary neural circuit affected in classical ADHD patients is the frontal-striatal circuit. They conclude that these findings indicate that different structures are related to the ADHD subtypes and also suggest that they may represent what are actually different phenotypes.

Finally, there is research to suggest that the presence or absence of tics and stereopathies can distinguish between ADHD combined type (ADHD-C) and Inattentive type presentations (Mariotti et al., 2021). Their results lent support to the hypothesis that the ADHD-C type includes abnormal involuntary movements in its phenomenology, which pointed to a broad spectrum of fronto-striatal dysfunction, of which tics, indicative of Tourette's disorder (Tourettes Plus), possibly is the complete expression. In this group, almost all the children evaluated (571/576) had involuntary motor movement of some type, and half had identifiable tic disorder. In addition, almost 30% of that population demonstrated stereopathies. Children with the inattentive type reported a much lower frequency of stereopathies (less than 10%) and did not exhibit tics at all. The authors concluded that the detection of tics and stereotypies could be a useful support to the diagnosis of ADHD. Furthermore, the absence of tics in the Inattentive subgroup was considered an additional element in favor of considering the Inattentive subgroup as a separate disorder instead of a subtype of ADHD.

These findings support the notion that the ADHD diagnostic construct may be too broad, encompassing what may actually be two discernible disorders with different underlying structural involvement as well as resultant functional presentation.

Genetic Evidence of Different ADHD Conditions

It is well established that attention-deficit/hyperactivity disorder (ADHD) is a complex neurodevelopmental disorder. As we have indicated, ADHD, as it is identified in clinical practice, is not an etiologically homogeneous disorder. ADHD almost certainly represents the common behaviorally manifested outcome of multiple causal pathways, both from a neural network perspective and from a genetic perspective. Research demonstrates that genetic factors make a significant contribution to many of these pathways (Coghill & Banaschewski, 2009). Behavioral studies suggest a heritability of approximately 0.76. Molecular genetic studies (Faraone et al., 2005; Thapar et al., 2006) have identified a range of potential candidate genes. From these studies, we have learned that the genetics of ADHD are characterized by a number of genes, each of which makes a small but significant contribution to the overall risk of the disorder becoming behaviorally manifest. Several regions have been supported across several studies (bin 16.4, 5p13, 11q22–25, 17p11) (Ogdie et al., 2004). The contribution of several candidate genes has been supported by meta-analyses (*DRD4*, *DRD5*, *DAT1*, *HTR1B*, and *SNAP25*) (Li et al., 2006). Recent research has expanded on this group of candidate genes (Demontis et al., 2023). They utilized a genome-wide association study meta-analysis of ADHD comprising 38,691 subjects with ADHD and 186,843 controls. The results indicated 27 genome-wide significant loci, highlighting 76 potential risk genes enriched among genes expressed particularly in early brain development. The genetic risk of ADHD was associated with several brain-specific neuronal subtypes and midbrain dopaminergic neurons. In exome-sequencing data (the exome, which is the part of the genome that codes for proteins; the exome makes up about 1% of the genome but contains around 85% of known disease-related variants) from 17,896 individuals, they identified an increased load of rare protein-truncating variants in ADHD for a set of risk genes enriched with probable causal common variants, potentially, implicating *SORCS3* in ADHD by both common and rare variants. They found 84–98% of ADHD-influencing variants were shared with other psychiatric disorders. In addition, common-variant ADHD risk was associated with impaired complex cognition such as verbal reasoning and a range of executive functions, including attention.

Even within the traditionally defined subdomains of ADHD, there is genetic variance that calls into question whether or not these subdomains should be grouped together under one general condition. Kunsti et al. (2014), based upon their review, conclude that "both shared and unique genetic risk factors underlie the two symptom domains of attention deficit hyperactivity disorder (ADHD): inattention and hyperactivity-impulsivity. The developmental course and relationship to co-occurring disorders differs across the two symptom domains, highlighting the importance of their partially distinct etiologies" (pg. 127).

Recently, different types of developmental trajectories have been identified within the ADHD population. These trajectories differ both in network properties and genetic contribution (Wang, 2024). Their research has identified three distinct trajectories: ascending, stable, and descending. They also examined the difference in the absolute value of the groups and labeled them ascending-high, stable-low, and descending-medium. Of note, specific attention hub regions provided the neural underpinnings for these three developmental trajectories. These regions encompassed the left anterior cingulate cortex/medial prefrontal cortex (ACC/mPFC), responsible for inhibitory control; the right inferior parietal lobule (IPL), which facilitated conscious focus on exogenous stimuli; and the bilateral middle frontal gyrus/precentral gyrus (MFG/PCG), responsible for regulating both dorsal and ventral attention networks. The MFG/PCG also played a central role in flexible modulation of endogenous and extrinsic attention. As to genetic contribution, individuals in the ascending-high group exhibited the highest Poly Genetic Risk Score (PRS) for ADHD, followed by those in the descending-medium group, with individuals in the stable-low group displaying the lowest PRS. Both ascending-high and descending-medium groups had significantly higher PRS compared to the stable-low group. Polygenic risk scores (PRS) are numeric estimates of an individual's genetic likelihood of developing a disease or trait at birth. PRSs are based on genome-wide genetic data, such as single-nucleotide polymorphisms (SNPs), protein-coding genes, or whole-genome sequencing (WGS). They are calculated by weighting risk alleles for different phenotypes by their effect size estimates and then adding them together (Wray et al., 2012). Age has no main effect on trajectories. The study identified attention hub regions as the neural basis of the aforementioned trajectories. They concluded that the identified group trajectories were driven by ADHD symptoms rather than emotional or conduct issues.

Environmental Contributions to Genetic Etiology in ADHD

Behavioral genetic investigations have consistently demonstrated large genetic influences for the core symptom dimensions of attention-deficit/hyperactivity disorder (ADHD), namely inattention and hyperactivity (Nigg et al., 2010). They conducted a meta-analysis to investigate etiological influences for inattention and hyperactive type by comparing the type and magnitude of genetic and environmental influences of each. Their results demonstrated that both symptoms were highly heritable (genetic factors accounted for 71% and 73% of the variance in inattentive and hyperactive types, respectively). However, the two sets of symptoms were dissociable as to the type of genetic influence. Dominant genetic effects were significantly larger for inattentive than for hyperactive, whereas additive genetic effects were larger for hyperactive than for inattentive type. Estimates of unique environmental effects were small to moderate and shared environmental effects were negligible for both symptom dimensions. The pattern of results generally persisted across several moderating factors, including gender, age, informant, and measurement method. The results suggested that inattentive type and hyperactive type may be linked to partially distinct neuropsychological mechanisms and temperament traits. There is research to support the contention that ADHD inattentive type is a distinct disorder from ADHD hyperactive and impulsive type (Grizenko et al., 2010; Milich

& Lynam, 2001). They found that the significant differences found between the hyperactive and the inattentive subtypes raise the possibility that the two may be separate disorders. Specifically, the children with the hyperactive subtype show both a higher frequency of conduct disorder and a good response to treatment. They are exposed to more moderate stress during their mothers' pregnancy and show a higher frequency of L/L genotype for the 5-HTT-linked polymorphic region than do children diagnosed with inattentive presentation. Additionally, subsequent research has supported these findings and further suggested that there may be genetic predispositions or potential susceptibility to environmental influences that may exacerbate or modulate this expression of the behavioral aspects of the disorder (Schwabe et al., 2024). Furthermore, there is evidence that indicates these environmental influences are more impactful earlier on in the development, and the phenotypical differences observed in adolescents and adults with ADHD are due to genetic variations (Hur, 2014). This data strongly suggests differing disorders that, using the current behavioral nosology, are not individually identified or studied. The above represents an ongoing issue with the heterogeneous nature of the current ADHD diagnostic nosology. The various, potentially significantly different disorders are lumped together for research purposes, thereby obscuring critical differences that may inform research and treatment.

ADHD and Functional Neuro-connectivity

There are also issues with this behavioral nosology system as it pertains to a network understanding of the disorder. The research on interconnectivity of networks and ADHD is in its infancy, but what there is supports the genetic findings of significant differences in subtypes. Specifically, neuroimaging findings in ADHD using diffusion tensor imaging (DTI) or resting-state functional MRI (rsfMRI) support the concept that a dysfunction in neurocircuitry may underlie this disorder we call ADHD and that differences in neural connectivity could distinguish between subtypes (Saad et al., 2020).

Some work on intrinsic networks has also focused on subnetworks associated with the behavioral, cognitive, and emotional symptoms related to ADHD. In particular, research focused on the default mode network (DMN) in ADHD for its role in self-referential thought, non-goal-directed activity, and "mind wandering" (Lee et al., 2021). Using diffusor tensor imaging, they assessed both structural and functional network efficiency and structure-functional network (SC-FC) coupling at the default mode (DMN), executive control (ECN), and salience (SAN) intrinsic networks in 201 children diagnosed with the inattentive subtype (ADHD-I) or the combined subtype (ADHD-C). They compared this data with data from typically developing children (TDC) to characterize ADHD symptoms relative to TDC and to assess differences between the various ADHD subtypes.

A note of importance before we get to the results: It is important to note that brain network theory posits that the structural network (white matter tracts) constrains the functional brain network (correlated functional activity), but functional connectivity also deviates from the observed structural architecture (Park & Friston, 2013). In general, however, structural connectivity in the brain is highly associated with functional connectivity. In the brain network topology, structural hub nodes tend to play central roles in functional networks (Shen et al., 2012). As mentioned, despite a significant overlap between structural and functional connectivity, functional connectivity is not fully restrained by the structural connectivity. One such estimate is that no more than 50% of the functional connectivity measured at a macroscopic scale corresponds to structural architecture in a strictly linear manner (Suarez et al., 2020). The structural connectivity and function connectivity are not uniformly related but gradually decouple according to the macroscale gradient. It is important to note that in the Lee et al. (2021) study, the structural properties differ more than the functional properties of networks between children with ADHD and TDC. Only structural connectivity and the efficiency of the structural network were found to be lower in children with ADHD compared to TDC.

Specifically, Lee et al.'s (2012) results indicated that relative to TDC, children with ADHD had lower structural connectivity and network efficiency in the DMN, without significant group differences in functional networks. Children with ADHD-C had higher SC-FC coupling, a finding consistent with diminished cognitive flexibility, for all subnetworks. "The ADHD-C group also demonstrated increased SC-FC coupling in the DMN compared to the ADHD-I group. The correlation between SC-FC coupling and hyperactivity scores was negative in the ADHD-I, but not in the ADHD-C group." They concluded that their results suggested the ADHD-C and ADHD-I subtypes likely differed with respect to their underlying neuronal connectivity and that the added dimensionality of hyperactivity may not explain this distinction compared to TDC. "The ADHD-C group also demonstrated increased SC-FC coupling in the DMN compared to the ADHD-I group. The correlation between SC-FC coupling and hyperactivity scores was negative in the ADHD-I, but not in the ADHD-C group." They conclude that ADHD-C and ADHD-I may differ with respect to their underlying neuronal connectivity.

Similarly, Saad et al. (2022) noted that ADHD-combined type (ADHD-C) was differentiated from ADHD-inattentive (ADHD-I) type by reduced network connectivity differences within the default mode network (DFN). Additionally, reduced white matter connectivity within the DMN was negatively associated with an ADHD rating scale hyperactivity-impulsivity subscale score. Compared with controls, ADHD-C but not ADHD-I differed by reduced connectivity within the DMN; inter-network connectivity between the DMN and somatomotor networks; the DMN and limbic networks; and between the somatomotor and cingulo-frontoparietal, with ventral attention and dorsal attention networks.

As a reminder, the default mode network (DMN) is a large-scale network of brain areas that are active when people are not engaged in a specific cognitive task. It's also known as the "task negative" network because its areas are less activated during tasks that require a lot of attention. The DMN is active during rest, passive tasks, and directed tasks that require remembering past events or imagining future events. It is also associated with internal processes like mind wandering, mental time travel, and perspective shifting.

In a study exploring the multimodal differences between the inattentive ADHD (ADHD-I) subtype and the combined ADHD (ADHD-C) subtype, Wu et al. (2022) utilized a battery of multiple scales and cognitive tests to assess the clinical and cognitive profiles of each individual. In addition, structural and diffusion magnetic resonance imaging (MRI) were obtained for 120 subjects with ADHD and 85 controls. Compared with controls, both ADHD groups demonstrated heightened levels of behavioral and emotional problems. The ADHD-C group had more behavioral problems and emotional lability, as well as less anxiety than the ADHD-I group. The two ADHD groups were equally impaired in most cognitive domains, with the exception of sustained attention. Compared with healthy controls, the ADHD-C group showed a high gray matter volume (GMV) in the bilateral thalamus and a high white matter volume in the body of the corpus callosum, while the ADHD-I group presented an elevated GMV mainly in the left precentral gyrus and posterior cingulate cortex. When compared with participants with ADHD-C and healthy controls, subjects with ADHD-I showed increased axial diffusivity in widespread brain regions. Overall, this study substantiated a distinct, interconnected pattern of behavioral, cognitive, and brain structural characteristics in children with different ADHD subtypes.

White Matter Disorder

Neurodevelopmental disorders (NDDs) usually present overlapping symptoms, which, on occasion, especially in young children (Mayes et al., 2012), make parsing these disorders into discrete entities difficult. Evidence has shown that NDDs have overlapping phenotypes, frequently co-occur, and share multiple genetic causes. Abnormal white matter (WM) microstructure has been found in many of these disorders (Zhao et al., 2022). Their research demonstrated that white matter microstructural abnormalities in the splenium of the corpus callosum were possibly shared among autism spectrum disorder (ASD) and attention deficit hyperactivity disorder (ADHD). In fact, up to 28% of individuals with ASD present comorbidity with ADHD (Jensen & Steinhausen, 2015). This data suggests that future research may be able to use white matter patterns to classify and regroup certain classes of disorders now considered to be disparate.

Furthermore, research has shown that ADHD subtypes might be differentiated by patterns of white matter dysfunction (Ercan et al., 2016). They applied tract-based spatial statistics to the Diffusor Tensor Imaging studies TI measures for obtaining fractional anisotropy (FA) and axial, radial diffusivity (AD, RD) measures to explore ADHD type-related differences in WM for the whole brain. Comparing ADHD-Combined group (ADHD-C) with the ADHD predominantly inattentive group (ADHD-I), they found "increased RD in several bilateral brain areas and increased AD mostly in left side of the brain, including the body and splenium of the corpus callosum, the anterior and posteriors limbs of the internal capsule, the superior, anterior and posterior corona radiata, the posterior thalamic radiation, and the superior longitudinal fasciculus." In the overlapping brain areas, the ADHD-C group presented increased AD values than did ADHD-I. They concluded that significant differences among ADHD types could be preliminary evidence (Saad et al., 2022) of distinct microstructural properties.

Emotional Dysregulation, Development, and ADHD

There is some research that suggests that the recognized childhood forms of ADHD do not persist into adulthood and that a better categorization of ADHD inclusive of emotionality is available. Adults with ADHD tend to display fewer hyperactive symptoms and more emotional symptoms as age increases (Barkley & Brown, 2008). Many patients with ADHD have emotional symptoms that are not part of the *DSM* criteria for the attention deficit hyperactivity disorder (Reimherr et al., 2020). Research has demonstrated that emotional symptoms are an important part of ADHD and that emotional lability makes a significant contribution to the poor functioning associated with behavioral impairment demonstrated in adults with ADHD (Barkley & Fischer, 2010; Skirrow & Asherson, 2013). In some cases, it is the emotional dysregulation that characterizes the prime disruptive feature of the disorder.

This emotionality in ADHD displays a definite genetic pattern (Biederman et al., 2012), with data linking it to hyperactivity/restlessness, impulsivity, inattention, and emotional dysregulation (Merwood et al., 2014).

Using a 2-factor, factor-analytic model, (1) attention difficulties and disorganization, and (2) emotional dysregulation, Reimherr et al. (2020) categorized subjects into two types based on levels of emotional symptoms. Individuals high in emotionality were labeled as having "ADHD emotional dysregulation presentation," and those low in emotionality were labeled as having "ADHD inattentive presentation." ADHD emotional dysregulation presentation subjects showed more childhood ADHD symptoms, adult symptoms of oppositional defiant disorder, and evidence of personality disorder, supporting their conclusion that ADHD in adults presents as two types and that adults with the emotional dysregulation presentation demonstrate higher levels of impairment and comorbidity. These results support the validity of the two diagnostic types in ADHD adults: inattentive presentation and emotional dysregulation presentation. This research is significant for its suggestion of a significant developmental trajectory for ADHD as well as clear variations into a final adult form.

There are other models that integrate white matter issues with emotional regulation issues in an attempt to compile a different cluster of ADHD disorders. For example, Karalunas et al. (2014) identified three distinct subtypes of ADHD based on profiles informed by temperament. These included a "Mild" type, whose individuals are characterized only by deficits in core ADHD symptom domains; a "Surgent" type, characterized by high levels of positive approach-motivated behaviors and activity level, shorter cardiac pre-ejection period, parasympathetic withdrawal in response to positive emotions, and atypical amygdala connectivity to medial frontal areas; and finally, they proposed an "Irritable" type, characterized by high levels of negative emotionality, weak parasympathetic response to negative emotional stimuli, reduced amygdala-insula connectivity, and a doubling of risk for onset of new behavioral or emotional disorders.

This developmental trajectory may be more complex than it seems. Longitudinal studies suggest the possibility of at least four developmental trajectories: early onset (3–5 years), middle childhood (6–14 years) onset with a persistent course, middle childhood onset with adolescent offset, and adolescent or adult onset (16 years and older) (Posner et al., 2020). Furthermore, ADHD symptomology may not be stable across the developmental span. Many children diagnosed with ADHD-C eventually transition to ADHD-I, given that the inattention symptom is relatively stable across development. In contrast, the hyperactivity/impulsivity symptoms often diminish with age. Accordingly, the DSM-5 changed the "type" designation used in the DSM-IV to a "presentation" terminology that hypothetically better captures the person's current symptomatology and its instability, in contrast to the more stable, trait-like characteristic denoted by the "type" terminology (de la Pena et al., 2020).

It appears from the above that the designation of attention deficit hyperactivity disorder represents a very "big tent," encompassing a myriad of forms and behavioral presentations. In addition, there are significant genetic, white matter, and environmental differences among its current constituent members. The current nosology recognizes three distinct processes, but there appear to be many more. Furthermore, there appears to be a case to be made for one of those three recognized disorders, ADHD-Inattentive Type, to be a completely separate disorder. It is likely a separate disorder representing attentional dysfunction, but it is probably not attention deficit hyperactivity disorder as we would like to identify it. That is the point. We are suggesting that there are lots of disorders of attention that are not ADHD, and by collapsing them into one, overarching diagnostic entity, we are potentially obscuring important and potentially clinically relevant information.

References

American Psychiatric Association. (1968). *Diagnostic and statistical manual of mental disorders (DSM-II)* (2nd ed.). American Psychiatric Association.

American Psychiatric Association (Ed.). (2013). *The diagnostic and statistical manual of mental disorders (DSM–5)* (5th ed.). American Psychiatric Association.

Barkley, R. A., & Brown, T. E. (2008). Unrecognized attention-deficit/hyperactivity disorder in adults presenting with other psychiatric disorders. *CNS Spectrums, 13*(11), 977–984.

Barkley, R., & Fischer, M. (2010). The unique contribution of emotional impulsiveness to impairment in major life activities in hyperactive children as adults. *Journal of the American Academy of Child and Adolescent Psychiatry, 49*(5), 503–513.

Biederman, J., Petty, C. R., Woodworth, K. Y., Lomedico, A., Hyder, L. L., & Faraone, S. V. (2012). Adult outcome of attention-deficit/hyperactivity disorder: A controlled 16-year follow-up study. *The Journal of Clinical Psychiatry, 73*(7), 577.

Bush, G. (2010). Attention-deficit/hyperactivity disorder and attention networks. *Neuropsychopharmacology Reviews, 35*, 278–300. https://doi.org/10.1038/npp.2009.120

Clark, L. C.-F. (2017, December 6). *Three approaches to understanding and classifying mental disorder: ICD-11, DSM-5, and the National Institute of Mental Health's Research Domain Criteria (RDoC).* https://doi.org/10.1177/1529100617727266. Retrieved from Sage Journals Psychological Science in the Public Interest.

Coghill, D., & Banaschewski, T. (2009). The genetics of attention-deficit/hyperactivity disorder. *Expert Review of Neurotherapeutics, 9*(10), 1547–1565. https://doi.org/10.1586/ern.09.78

de la Pena, M., Pan, M., Thai, C., & Alisso, T. (2020). Attention-deficit/hyperactivity disorder predominantly inattentive subtype/presentation: Research progress and translational studies. *Brain Science, 10*(5), 292. https://doi.org/10.3390/brainsci10050292

Demontis, D., Walters, G. B., Athanasiadis, G., Walters, R., Therrien, K., Nielsen, T. T., ... & iPSYCH-Broad Consortium. (2023). Genome-wide analyses of attention deficit hyperactivity disorder identify 27 risk loci, refine the genetic architecture, and implicate several cognitive domains. *Nature Genetics, 55*(2), 198.

Diamond, A. (2005). Attention-deficit disorder (attention-deficit/hyperactivity disorder without hyperactivity): A neurobiologically and behaviorally distinct disorder from attention-deficit/hyperactivity disorder (with hyperactivity). *Development and Psychopathology, 17*(3), 807–825.

Ercan, E. S., Suren, S., Bacanlı, A., Yazıcı, K. U., Callı, C., Ardic, U. A., ... & Rohde, L. A. (2016). Altered structural connectivity is related to attention deficit/hyperactivity subtypes: A DTI study. *Psychiatry Research: Neuroimaging, 256*, 57–64.

Faraone, S. V., Perlis, R. H., Doyle, A. E., Smoller, J. W., Goralnick, J. J., Holmgren, M. A., & Sklar, P. (2005). Molecular genetics of attention-deficit/hyperactivity disorder. *Biological Psychiatry, 57*(11), 1313–1323.

Girault, J., Veatch, O., & Won, H. (2024). Etiologic heterogeneity, pleiotropy, and polygenicity in behaviorally defined intellectual and developmental disabilities. *Journal of Neurodevelopmental Disorders, 16*, 8. https://doi.org/10.1186/s11689-024-09526-z

Grizenko, N., Paci, M., & Joober, R. (2010). Is the inattentive subtype of ADHD different from the combined/hyperactive subtype? *Journal of Attention Disorders, 13*(6), 649–657. https://doi.org/10.1177/1087054709347200

Hur, J. (2014). Dimensions of psychopathology and trait affect do not predict noticing of an unexpected non-emotional object (Doctoral dissertation, University of Illinois at Urbana-Champaign).

Jensen, C., & Steinhausen, H. (2015). Comorbid mental disorders in children and adolescents with attention-deficit/hyperactivity disorder in a large nationwide study. *ADHD Attention Deficit and Hyperactivity Disorders, 7*(1), 27–38.

Karalunas, S. L., Fair, D., Musser, E. D., Aykes, K., Iyer, S. P., & Nigg, J. T. (2014). Subtyping attention-deficit/hyperactivity disorder using temperament dimensions: Toward biologically based nosologic criteria. *JAMA Psychiatry, 71*(9), 1015–1024.

Kessler, R. C., et al. (2006). The prevalence and correlates of adult ADHD in the United States: Results from the National Comorbidity Survey Replication. *The American Journal of Psychiatry, 163*(4), 716–723.

Kunsti, J., et al. (2014). The separation of ADHD inattention and hyperactivity-impulsivity symptoms: Pathways from genetic effects to cognitive impairments and symptoms. *Journal of Abnormal Child Psychology*, 127–136. https://doi.org/10.1007/s10802-013-9771-7

Lee, P. C., Niew, W. I., Yang, H. J., Chen, V. C. H., & Lin, K. C. (2012). A meta-analysis of behavioral parent training for children with attention deficit hyperactivity disorder. *Research in Developmental Disabilities, 33*(6), 2040–2049.

Lee, D., Knight, E. Q., Song, H., Lee, S., Pae, C., & Yoo, S. (2021). Differential structure-function network coupling in the inattentive and combined types of attention deficit hyperactivity disorder. *PLoS One*. https://doi.org/10.1371/journal.pone.0260295

Li, D., Sham, P. C., Owen, M. J., & He, L. (2006). Meta-analysis shows significant association between dopamine system genes and attention deficit hyperactivity disorder (ADHD). *Human Molecular Genetics, 15*(14), 2276–2284.

Magnus, W. N. (2023, August 8). *Attention deficit hyperactivity disorder*. Retrieved from National Library of Medicine, National Center for Biotechnology Information. https://www.ncbi.nlm.nih.gov/books/NBK441838/

Mariotti, P., Mariotti, P., Lucibello, S., De Clemente, V., Veltri, S., Janiri, D., et al. (2021). ADHD syndrome and tic disorders: Two sides of the same coin. *The Lancet*. https://doi.org/10.2139/ssrn/3923340

Mayes, S., Calhoun, S., Mayes, R., & Molitoris, S. (2012). Autism and ADHD: Overlapping and discriminating symptoms. *Research in Autism Spectrum Disorders, 6*(1), 277–285. https://doi.org/10.1016/j.rasd.2011.05.009

Merwood, A., Chen, W., Rijsdijk, F., Skirrow, C., Larsson, H., Thapar, A., ... & Asherson, P. (2014). Genetic associations between the symptoms of attention-deficit/hyperactivity disorder and emotional lability in child and adolescent twins. *Journal of the American Academy of Child & Adolescent Psychiatry, 53*(2), 209–220.

Milich, R. B., & Lynam, D. (2001). ADHD combined type and ADHD predominantly inattentive type are distinct and unrelated disorders. *Clinical Psychology: Science and Practice, 8*(4), 463–488. https://doi.org/10.1093/clipsy.8.4.463

Mu, S., Wu, H., Zhang, J., & Chang, C. (2022). Structural brain changes and associated symptoms of ADHD subtypes in children. *Cerebral Cortex, 32*(6), 1152–1158.

Nigg, J., Nikolas, M., & Burt, S. A. (2010). Measured gene-by-environment interaction in relation to attention-deficit/hyperactivity disorder. *Journal of the American Academy of Child & Adolescent Psychiatry, 49*(9), 863–873.

Ogdie, M. N., Fisher, S. E., Yang, M., Ishii, J., Francks, C., Loo, S. K., ... & Nelson, S. F. (2004). Attention deficit hyperactivity disorder: fine mapping supports linkage to 5p13, 6q12, 16p13, and 17p11. *The American Journal of Human Genetics, 75*(4), 661–668.

Park, H., & Friston, K. (2013). Structural and functional brain networks: From connections to cognition. *Science, 342*(6158), 1238411. pmid:24179229.

Posner, J., Polanczyk, G., & Sonuga-Barke, E. (2020). Attention-deficit hyperactivity disorder. *Lancet, 395*(10), 450–462. https://doi.org/10.1016/S0140-6736(19)33004-1

Reimherr, F., Roesler, M., Marchant, B., Gift, T., Tetz, W., & Philipp-Wiegmann, F. (2020). Types of adult attention-deficit/hyperactivity disorder: A replication analysis. *The Journal of Clinical Psychiatry, 81*(2). https://doi.org/10.4088/JCP.19m13077

Rosales, A., Vitoratou, S., Banaschewski, T., Asherson, P., Buitelaar, J., Oades, R., et al. (2015). Are all the 18 DSM-IV and DSM-5 criteria equally useful for diagnosing ADHD and predicting comorbid conduct problems? *European Child Adolescent Psychiatry, 24*, 1325–1337. https://doi.org/10.1007/s00787-015-0683-7

Saad, J. F., Griffiths, K. R., & Korgaonkar, M. S. (2020). A systematic review of imaging studies in the combined and inattentive subtypes of attention deficit hyperactivity disorder. *Frontiers in Integrative Neuroscience, 14*, 31.

Saad, J., et al. (2022). Intrinsic functional connectivity in the default mode network differentiates the combined and inattentive attention deficit hyperactivity disorder types. *Frontiers in Human Neuroscience, 16*. https://doi.org/10.3389/fnhum.2022.859538

Schwabe, I., Jović, M., Rimfeld, K., Allegrini, A. G., & van den Berg, S. M. (2024). Genotype-environment interaction in ADHD: Genetic predisposition determines the extent to which environmental influences explain variability in the symptom dimensions hyperactivity and inattention. *Behavior Genetics, 54*(2), 169–180.

Semrud-Clikeman, M., Fine, J. G., Bledsoe, J., & Zhu, D. C. (2017). Regional volumetric differences based on structural MRI in children with two subtypes of ADHD and controls. *Journal of Attention Disorders, 21*(12), 1040–1049.

Shen, K., et al. (2012). Information processing architecture of functionally defined clusters in the macaque cortex. *Journal of Neuroscience, 32*(48), 17465–17476. Epub 2012/12/01. pmid:23197737.

Skirrow, C., & Asherson, P. (2013). Emotional lability, comorbidity and impairment in adults with attention-deficit hyperactivity disorder. *Journal of Affective Disorders, 147*(1–3), 273–281.

Suarez, L., Markello, R., Betzel, R., & Misic, B. (2020). Linking structure and function in macroscale brain networks. *Trends in Cognitive Sciences, 24*(4), 302–315. Epub 2020/03/12. pmid:32160567.

Thapar, A., Langley, K., O'Donovan, M., & Owen, M. (2006). Refining the attention deficit hyperactivity disorder phenotype for molecular genetic studies. *Molecular Psychiatry, 11*(8), 714–720.

Wang, Y. M. (2024). The neural and genetic underpinnings of different developmental trajectories of Attention-Deficit/Hyperactivity Symptoms in children and adolescent. *BMC Medicine*. https://doi.org/10.1186/s12916-024-03449-1

Wray, N. R., Pergadia, M. L., Blackwood, D. H. R., Penninx, B. W. J. H., Gordon, S. D., Nyholt, D. R., ... & Sullivan, P. F. (2012). Genome-wide association study of major depressive disorder: New results, meta-analysis, and lessons learned. *Molecular Psychiatry, 17*(1), 36–48.

Yadav, S. K., Bhat, A. A., Hashem, S., Nisar, S., Kamal, M., Syed, N., ... & Haris, M. (2021). Genetic variations influence brain changes in patients with attention-deficit hyperactivity disorder. *Translational Psychiatry, 11*(1), 349.

Zhao, Y., Yang, L., Gong, G., Cao, Q., & Liu, J. (2022). Identify aberrant white matter microstructure in ASD, ADHD and other neurodevelopmental disorders: A meta-analysis of diffusion tensor imaging studies. *Progress in Neuro-Psychopharmacology and Biological Psychiatry, 113*, 110477. https://doi.org/10.1016/j.pnpbp.2021.110477

Chapter 5
Neural Networks of Attention

Attention is a construct we all refer to with casual ease. However, the construct has been under study for a long time in order to better define the process and its components. Neuroimaging is allowing us to better connect the neural network system with this construct. It is also allowing us to better elucidate and functionally map different aspects of that construct we call attention in order to validate or negate both theoretical and historical constructs.

Attention Network Theory (Posner & Petersen, 1990; Petersen & Posner, 2012) hypothesizes three largely independent neural network systems that underpin each of three different types (functions) of attention. These dissociable networks are the alerting system, which initiates a state of increased arousal in direct anticipation of upcoming stimuli; the orienting system which shifts the attentional focus to locations in space; and the control system, which selects and emphases the most relevant information when distracting or task-incompatible information is present. The three systems were thought to dissociate neuroanatomically into independent "attention networks" (Markett et al., 2022). It is important to note that, while these systems were considered anatomically dissociable, they were posited to be heavily integrated on a functional level.

The Basics of Neural Network Formation and Operation

Let us turn briefly to an updating review of the hierarchical structure of neural networks. Researchers have historically attributed independent encoding functions to the activity of single neurons (Evarts, 1868) or, somewhat later, groups of neurons (Humphrey & Thompson, 1970). That is, the neuron was envisioned as the fundamental computational unit of the brain. The focus of this research was to identify correlations between specific neural activity and specific movement or cognitive

T. Wasserman, L. D. Wasserman, *Disorders of Attention*, Neural Network Model: Applications and Implications, https://doi.org/10.1007/978-3-031-78732-4_5

activities. The belief was that neural activity would encode particular activities in a consistent and highly predictable manner.

Later research identified some holes in this line of reasoning. For example, research demonstrated considerable variability in the activity patterns across neurons (Churchland & Shenoy, 2007) and instability of movement parameter encoding by single neurons across different conditions (Scott & Kalaska, 1997). Later research further clarified the relationship of single neurons to the neural populations that they were embedded in. This research demonstrated long-term stability in single-neuron activity patterns across various brain regions. A small amount of drift in neural activity did occur and is explainable by concomitant changes in task-irrelevant aspects of the behavior. These results suggest that long-term stable behaviors are generated by single neuron activity patterns that are themselves highly stable (Jensen et al., 2022). This later research indicates that the same neuron can be recruited to perform its task as part of differing networks that are spontaneously produced to perform task-specific activities. As we shall see, this organizational property of the brain has important implications for attention.

More recent research has not been based on the independent modulation of single neurons but rather activities performed at the population level by networks of interconnected neurons. This idea leads to what has been termed the neural population doctrine. The neural population doctrine is the idea that the population of neurons, rather than the individual neuron, is the fundamental unit of computation in the brain. Populations of cortical neurons flexibly perform different functions. For example, with respect to the primary motor cortex, the population of neurons is involved in a rich repertoire of motor behaviors (Gallego et al., 2018). While each specific motor-involved task requires different patterns of muscle and single-unit activity, there are considerable similarities at the neural population level among the areas involved for different tasks. In sum, research has demonstrated that the structure and activity of the neural nodes are largely preserved across tasks.

Newer research provides further clarification of this process. For example, there is evidence indicating that the prefrontal cortex (PFC) constitutes the structural basis underlying cognitive flexibility for adaptive behavior. In support of these efforts, PFC neurons are not feature-specific. Rather, they demonstrate mixed selectivity and context-dependent coding of task features, including environmental factors such as sensory stimuli or task requirements such as abstract rules. Secondly, these mixed- selective neurons participate in multiple cognitive operations through transient coalitions of cell assemblies. Therefore, the process of information encoded by mixed-selective neurons is understood at the level of population activity. The PFC resolves cognitive interference between competing features by transforming their representational population geometry into distinct neural subspaces to accommodate flexible task-switching. These results elucidate a fundamental neural coding principle that constitutes a central building block of human flexible cognitive control (Weber et al., 2023). These subspaces are referred to as neural manifolds. A neural manifold is a low-dimensional subspace that's embedded in a high-dimensional neural state space and is explored by a specific neural populations' activity. The manifold's topological structure is formed by neural activity dynamics that often move through a lower-dimensional space. Neural manifolds can be found in sensory, motor, and cognitive brain regions. Furthermore, the activities of neurons within these manifolds are nonlinear. Diverse single-neuron responses can be well described by relatively few patterns of neural co-modulation. These neurons make thousands of distributed and recurrent connections that likely amplify nonlinearity. As a result, the operation of neural manifolds is nonlinear as well (Fortunato et al., 2024). Understanding the nonlinear operation of these manifolds will be critical to understanding how the human brain regulates complex behavior such as attention. As this relates to attention in specific, the population doctrine further suggests that these rapid switches between distinct cognitive operations are achieved efficiently by adjusting the population geometry, i.e., the reconfiguration of cell assemblies. A key advantage of this model is that neural representations that evolve in parallel can be integrated into a unified, connected representation (Weber et al., 2023). Hypothetically, a deficiency in switching efficiency might lead to some forms of attentional dysfunction.

What Is the Neural Population Level?

The brain is populated by millions of neurons that are organized in different brain areas. Within any specific brain area are different subregions. Each of these subregions is composed of different layers, and inside each layer, there are various cell types. While much research is focused on the operation of a single neuron, these populations of neurons at the subregion layer level (neuronal populations) have characteristic properties that are derived from their working together. Theory suggests a relationship between neuronal responses and psychophysical decisions, which may prove useful for identifying cell populations underlying specific perceptual capacities (Zohary et al., 1994). That allows for modeling that includes the unique response properties of these specific populations. For example, rather than looking at the response time of a single neuron which belongs to a certain cell class at a certain level, a number of questions might be reasonably asked. So, for example, supposing a human brain receives a certain type of stimulus, what is the activity of all the cells in a particular layer of the subregion processing the response, what is the activity of the population class of cells in that layer of the subregion, and finally, what is the response of this subregion as a whole? This, of course, leads to the question, what is the response of a brain area? In other words, at any of the scales of spatial resolution, it is productive to look at the response of the neuronal population as a whole rather than in the activities of individual neurons (Gerstner et al., 2014). As we have seen, the human brain creates some of these subpopulations (manifolds) of neurons on a task-specific basis (see Fig. 5.1). Different but related tasks likely require the recruitment of similar neuronal populations for their solution.

Hypothesis: varied motor behaviors are caused by the flexible activation of combinations of neural modes. (**a**) The connectivity of the cortical network results in neural modes whose combined activity explains the specific activity of individual

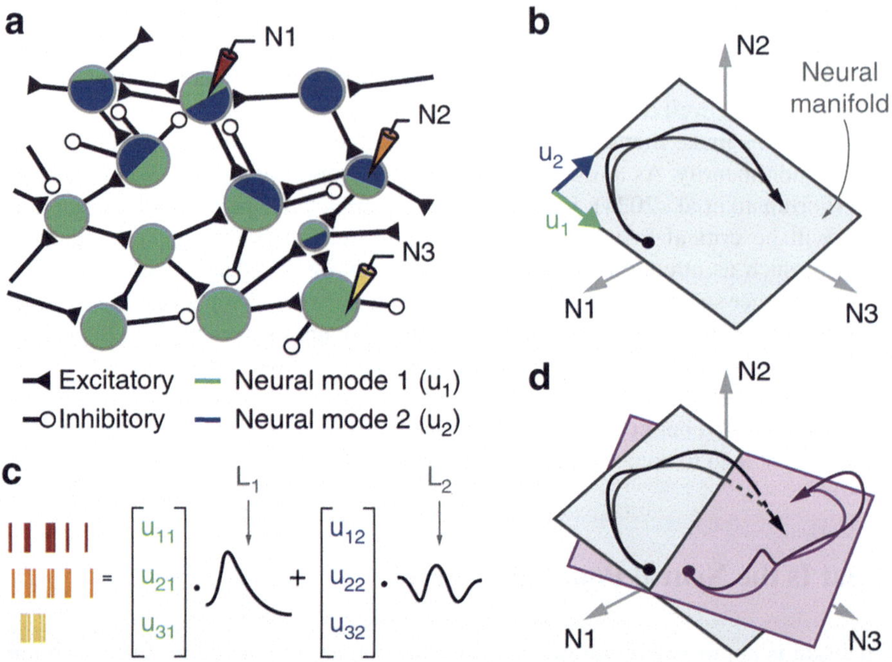

Fig. 5.1 (**a-d**) Cortical population activity within a preserved neural manifold underlies multiple motor behaviors

neurons. (**b**) The neural space for the three neurons recorded in **a**. The time-dependent population activity is represented by a trajectory (in black, the arrow indicates time direction) mostly confined to a two-dimensional neural manifold (gray plane) spanned by two neural modes (green u_1 and blue u_2 basis vectors) (Gallego et al., 2018)

Since each of these population centers participates in multiple types of motor responses, damage to any one of them impacts all of the downstream activities emanating from them. This implies that an individual with attention problems will likely experience potentially numerous disruptions in behavior downstream from the area of the neuronal population that is dysfunctional.

Integrating Neural Populations into a Coherent Neural Network

Now that we understand that there are groups of neurons that essentially operate together, in a nonlinear fashion, and as part of a larger network (composed of different manifolds), it is important to understand how these neuronal populations are related to the well-established hub and node modeling that is critical for understanding how neural networks operate.

Neural Hubs and Nodes

Effectively, every domain of cognitive function in the human brain requires the integration of distributed neural activity. Network analysis of human brain connectivity has routinely identified sets of regions that are essential for enabling efficient neuronal signaling and communication. These areas have been termed brain hubs or nodes. The central embedding of these candidate "brain hubs" in anatomical networks supports their diverse functional roles across a broad range of cognitive tasks and widespread dynamic coupling within and across functional networks. The high level of centrality and importance of these brain hubs also makes them points of vulnerability that are susceptible to disconnection and dysfunction in brain disorders (van den Heuvel & Sporns, 2013; Stam, 2024). The relationship between a neural node, neural population, and neural hub can be viewed as follows: A neural node can be as small as a single neuron and is a network component that has one or more input connections, a transfer function that combines the inputs in some way, and an output connection. In the context of a network, a hub is a node with a large number of connections with many other nodes. A neural population is essentially a highly integrated complex hub consisting of a large number of individual neurons operating in a synchronized manner. As regards our current discussion, damage or disruption of any of these components will impact the network of which they are a part.

There are a large number of terms that are associated with this modeling, and we are providing the reader with a glossary of these terms. The core of this glossary was generated by van den Heuvel and Sporns (2013).

Glossary (See Fig. 5.2)

Brain connectivity:	Description of structural or functional connectivity between brain network elements (i.e., brain regions, neurons).
Centrality:	Measures of the relative importance of a node or edge within the overall architecture of a network. Several centrality metrics have been proposed, including (among many others) degree, betweenness, closeness, eigenvector, and PageRank centrality.
Clustering:	The tendency of small groups of nodes to form connected triangles. Many triangles around a central node imply that the connected neighbors of the node are also neighbors of each other, forming a cluster or clique.
Community:	In networks, communities refer to modules, densely interconnected sets of nodes.

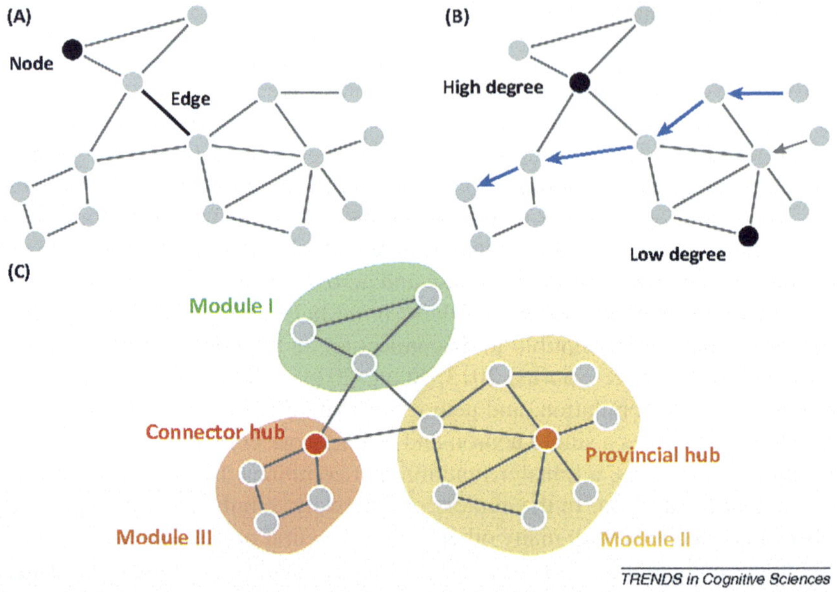

Fig. 5.2 Basic network attributes. (**a**) Brain networks can be described and analyzed as graphs comprising a collection of nodes (describing neurons/brain regions) and a collection of edges (describing structural connections or functional relationships). The arrangement of nodes and edges defines the topological organization of the network. (**b**) A path corresponds to a sequence of unique edges that are crossed when traveling between two nodes in the network. Low-degree nodes are nodes that have a relatively low number of edges; high-degree nodes (often referred to as hubs) are nodes that have a relatively high number of edges. (**c**) A module includes a subset of nodes of the network that show a relatively high level of within-module connectivity and a relatively low level of intermodule connectivity. "Provincial hubs" are high-degree nodes that primarily connect to nodes in the same module. "Connector hubs" are high-degree nodes that show a diverse connectivity profile by connecting to several different modules within the network. (Published in *Trends in Cognitive Sciences*, 2013: "Network Hubs in the Human Brain" by M. V. D. Heuvel, O. Sporns)

Connection matrix: A summary of all pairwise associations (connections) between network nodes, rendered in the form of a square matrix.

Connection: A connection expresses the existence and/or strength of a relationship, interaction, or dependency between two nodes in the network. Connections can be binary or weighted, and they can be directed or undirected. Connections are also referred to as edges.

Connectome: A comprehensive network map of the anatomical connections of a species' nervous system.

Connector hub: A high-degree network node that displays a diverse connectivity profile across several different modules in a network.

Core:	A group of nodes that share a large number of mutual connections, rendering them resistant to damage. Cores are identified by using a recursive procedure that prunes away weakly connected nodes.
Degree:	The number of edges attached to a given node.
Directed network:	A network comprising directed connections (edges).
Edge:	A term for a network connection.
Functional connectivity:	Measured as the statistical dependence between the time series of two network nodes (e.g., brain regions, neurons).
Graph:	A mathematical description of a network, comprising a collection of nodes and a collection of edges.
Hub:	A node occupying a central position in the overall organization of a network.
Manifold:	A neural manifold is a low-dimensional subspace that's embedded in a high-dimensional neural state space and is understood by a specific neural population's activity. The manifold's topological structure is formed by neural activity dynamics that often move through a lower-dimensional space.
Module:	A group of nodes that maintain a large number of mutual connections and a small number of connections to nodes outside their module.
Parcellation:	A subdivision of the brain into anatomically or functionally distinct areas or regions.
Participation coefficient:	A graph-theoretical measure that expresses the distribution of edges of a node across all modules in a network.
Provincial hub:	A high-degree network node that mostly connects to nodes within its own module.
Resting-state network:	A set of brain regions that show coherent functional connectivity during task-free spontaneous brain activity.
Rich-club organization:	The propensity of a set of high-degree nodes in a network to be more densely interconnected than expected on the basis of their node degree alone.
Scale-free organization:	A network with a degree distribution that follows a power-law function.
Shortest path length:	The shortest path length between two nodes reflects the minimal number of links that have to be crossed to travel from one node to another node in the network.
Small-world organization:	A network that shows a level of clustering higher than that observed in random networks and an average shortest path length that is equal to that observed in random networks.

Structural connectivity: A description of the anatomical connections between network nodes (i.e., brain regions, neurons); for example, reconstructed anatomical projections derived from diffusion MRI, directed anatomical pathways derived from neural tract tracing, or synaptic connections between individual neurons.

Undirected network: A network comprising undirected connections (edges).

What Is the Neural Substrate that Enables Integration of Distributed Neural Information and Thus the Emergence of Coherent Mental and Cognitive States?

Two aspects of brain organization appear essential in the understanding of how the brain generates complex action or thought. First, integration depends on neural communication among specialized brain regions, unfolding within a network of interregional projections (Fuster, 1997), which gives rise to large-scale patterns of coordinated activity (Singer, 1993) between connected elements of the network. Second, important integrative functions are performed by a specific set of brain regions and their anatomical connections. These regions are capable of complex and diverse responses and represent focal points of convergence or divergence of more specialized neural information ("confluence zones") (Meyer & Damasio, 2009). We have seen how the understanding of manifolds helps elucidate how this integration takes place.

Very recently, research has suggested that some of this integration happens at the subcortical level. For example, subcortical visual areas such as the dorsal lateral geniculate nucleus (dLGN) or the superior colliculus (SC) have long been held as basic structures responsible for a stable, integrated, and defined function. The dLGN can be considered a relay for visual information traveling from the retina to cortical areas and the SC as a sensory integrator orienting body movements toward visual targets. Findings suggest that both dLGN and SC neurons express functional plasticity, adding unexplored layers of complexity to their previously attributed functions (Duménieu et al., 2021).

The Human Attention System

We will now turn our focus to the human attentional system in specific. We will first review the traditionally hypothesized networks and their components. We will then discuss more recent brain mapping research that indicates that all attention domains recruited multiple and partly overlapping intrinsic networks and converged in the dorsal frontoparietal and midcingulate-insular networks (Markett et al., 2022). While this work highlighted a preference of each attentional function for its own set of intrinsic networks, the various mapped networks did not match well to those proposed in the traditional literature, suggesting that a necessary refinement of the attention network theory might be necessary. Specifically, the idea that these networks are dissociable and independent is incorrect. Instead, it is likely that attention is supported by a complex distributed network in which different subroutines of attention (alerting, orienting, and control) segregate into different subnetworks and are integrated by hubs in the dorsal frontoparietal and midcingulate-insular networks.

The Attention System of the Human Brain Traditional View

Posner and Petersen (1990) and later Petersen and Posner (2012) originally proposed three networks to encompass attentional functioning. These were an alerting network, which focused on brain stem arousal systems along with right hemisphere systems related to sustained vigilance; an orienting network focused on, among other regions, parietal cortex; and an executive network, which included midline frontal/anterior cingulate cortex. The original three were later added to, or elaborated upon, to include a self-regulation component. In this volume, we will also refer to these contributory networks as subroutines.

General Anatomy Components of the Attentional System

The general anatomy of the attention system was described as being contained in the dorsal visual pathway. This pathway has its primary cortical projection area in and around the calcarine fissure in the occipital lobe. This area bi-hemispherically receives information directly from its ipsilateral lateral geniculate nucleus that receives signals from the contralateral visual hemifield and extends into the parietal lobe (Petersen et al., 1988). The posterior attention system is thought to operate upon a ventral pathway during tasks requiring detailed processing of objects. This visual pattern recognition system is described as involving a ventral pathway, stretching from the calcarine fissure in the occipital lobe to the infratemporal cortex.

Expanding on this idea, there was seminal research that described the anatomical and functional organization of attentional systems in the human brain and distinguished these two distinct yet interacting frontoparietal networks: the dorsal network guiding the voluntary allocation of attention, and the ventral network responding to unexpected but relevant stimuli (Corbetta & Shulman, 2002, 2011).

Human Dorso-ventral Attentional Connection Identified Using Tractography (See Fig. 5.3)

Alerting

Alerting refers to an alert state that an individual attains and then maintains to deal with external inputs; it is associated with the locus coeruleus norepinephrine (LC-NE) system (Posner & Petersen, 1990). There is consistent evidence that associates norepinephrine (NE)-based networks with the alerting system. According to Peterson and Posner, a "warning signal" is accompanied by activity in the locus coeruleus (LC), which is a principal source of NE (Holland et al., 2021). NE participates in a variety of behavioral and physiological processes such as arousal, learning and memory, attention, mood, appetite, and stress reactivity (Saboory et al., 2020), altering cognitive function and performance in attentional tasks. The NE

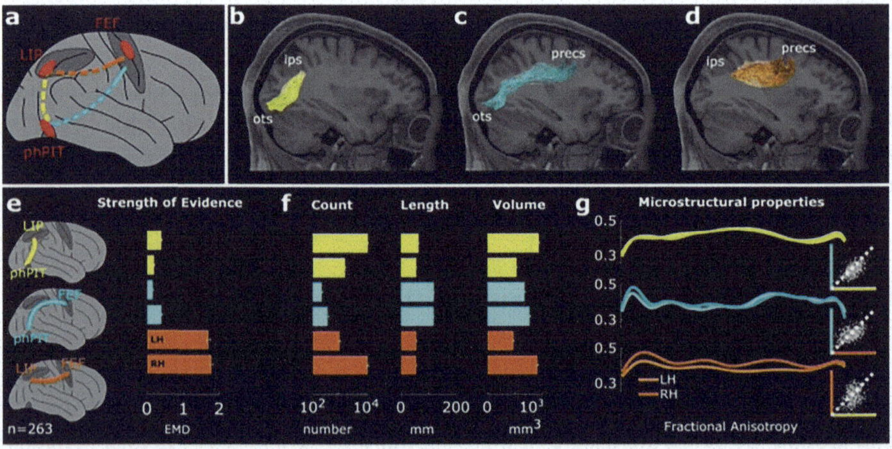

Fig. 5.3 (**a**) Schematic connections; phPIT–LIP: yellow trace; phPIT–FEF: cyan trace; LIP–FEF: orange trace. (**b–d**) Sagittal-view of phPIT-to-LIP, phPIT-to-FEF, and LIP-to-FEF connections overlaid on T1 image for subject 101006. Conventions as in Fig. 5.1. (**e**) The bar plot shows the mean earth mover's distance (EMD; see also Methods) in support of the existence of the tracts; lower bars for each tract represent the average across the right hemisphere, upper bars across the left. (**f**) Bar plots show the average streamline number, tract length (mm), and tract volume (mm^3) for functional tracts of 263 subjects; lower bars for each tract represent the average across the right hemisphere, and upper bars across the left. (**g**) Microstructural properties of functional tracts as measured by fractional anisotropy (FA). Each line represents the FA value averaged across subjects and calculated along the tract. Insets show direct comparison of individual subject FA between phPIT–LIP and phPIT–FEF connections (top), LIP–FEF and phPIT–FEF connections (middle), and phPIT–LIP and LIP–FEF connections (bottom). Data are expressed as mean across 263 subjects and separately for the two hemispheres; gray points represent the values for each individual subject. ips intraparietal sulcus, ots occipitotemporal sulcus, sts superior temporal sulcus, preces pre-central sulcus, FEF frontal eye field, LIP lateral intraparietal area, phPIT putative human posterior inferotemporal area, LH left hemisphere, RH right hemisphere (Sani et al., 2021)

pathway includes major nodes in the frontal cortex and in parietal areas relating to the dorsal but not the ventral visual pathways. The norepinephrine-based network has extensive connections cortically and subcortical. The projections of the LC are small unmyelinated fibers, forming a wide antero-posterior branching network to reach the raphe nuclei, the cerebellum, and almost all areas of the midbrain and forebrain regions with very few exceptions (hypothalamus and striatum) (Berger et al., 2021).

Interestingly, NE participates in the shaping and the wiring of the nervous system during the critical periods of development, and perturbations in this process can alter the brain's developmental trajectory, which in turn can cause long-lasting and even permanent changes in the brain function and behavior later in life. As it will pertain to a later discussion, we would also point out that exposure to stress, prenatal or early postnatal malnutrition, leads to a reprogramming of the brain noradrenergic signaling. We will return to this finding in our discussion of attention being a trans-diagnostic factor in many emotional dysregulation disorders.

Orienting Network

The orienting network is focused on the ability to prioritize sensory input by selecting a modality or location. Originally, both the pulvinar and the superior colliculus were described as essential for visual selection. Additionally, the parietal cortex was identified as part of a posterior attention system. That view was later modified to include the frontal (including the frontal eye fields) as well as posterior areas as part of the orienting process. Both dorsal and ventral orienting networks were described. The dorsal attention network consists of frontal eye fields (FEF) and the intraparietal sulcus/superior parietal lobe (IPS/SPL). The ventral attention network (teal) consists of regions in the temporoparietal junction (TPJ) and the ventral frontal cortex (VFC).

Executive Network

This set of processes is associated with the limited capacity of the attention system and the necessity for prioritizing objects for attention and future action. It was hypothesized as being related to awareness itself and has often been referred to as focal attention. This was originally described as involving widespread connections from the midline cortex and the anterior cingulate cortex (ACC). Focal attention is associated with activity found in the medial frontal/anterior cingulate in such diverse operations as perception of either physical or social pain, processing of reward, monitoring or resolution of conflict, error detection, and theory of mind. These different demands all activate this region, in most cases in conjunction with the anterior insula (Petersen & Posner, 2012).

Later research indicated that there were two separable executive control networks. The first, a frontoparietal network, is likely distinct from the orienting network, whereas the second, a cingulo-opercular network, overlaps with the original executive network. These networks were deemed dissociable, although developmentally, they both have a common early developmental precursor (Dosenbach et al., 2008).

These original networks were later added to by the addition of self-regulation. In adults, the same set of voluntary functions is frequently called self-control. In all instances, self-control or self-regulation was conjectured to be an ability to control reflexive or otherwise dominant responses to select less dominant ones. This network was described as including the anterior cingulate, anterior insula, and areas of the prefrontal cortex when inhibition of dominant responses was required. The common involvement of the anterior cingulate in attention and both emotion and cognitive control provided the basis for the argument that the executive attention network is critical to these various functions (Petersen & Posner, 2012).

As we have discussed, Posner's model was not the only model of attention. Other models described similar discriminating functions, using slightly different and, in some cases, expanded nomenclature. Certainly, on a behavioral level, the relative independence of these activities has been amply described in the literature, but as we shall now see, that is not the entire story.

Brain Mapping and Attentional Domains

Although the independence of the functions described in the Network Attention Theory has been demonstrated by research on the functional properties of attention, the neural mechanisms underlying their interactions have only recently been investigated. Using functional magnetic resonance imaging, the cortical and subcortical activity related to these attentional functions and their interactions have been studied (Xuan et al., 2016). Their results clarified that areas in the extended frontoparietal network (FPN), including dorsolateral prefrontal cortex, frontal eye fields (FEF), areas near and along the intraparietal sulcus, anterior cingulate and anterior insular cortices, basal ganglia, and thalamus, were activated across multiple attentional functions. Specifically, the alerting function was associated with activation in the locus coeruleus (LC) in addition to regions in the FPN. The orienting functions were associated with activation in the superior colliculus (SC) and the FEF. The executive control function was mainly associated with activation of the FPN and cerebellum. The interaction effect of alerting by executive control was also associated with the activation of the FPN, while the interaction effect of orienting validity by executive control was mainly associated with the activation in the pulvinar. The current findings demonstrate that cortical and specific subcortical areas play a pivotal role in the implementation of the range of attentional functions and underlie and contribute to their dynamic interaction. This conclusion was buttressed by research findings indicating that attentional control spans three cortical lobes and overarches large distances through fiber pathways that run orthogonally to the dominant anterior-posterior axes of sensory processing (Sani et al., 2021). Their research also showed that, similar to the frontoparietal attentional areas, the human posterior inferotemporal cortex exhibits significant attentional modulatory activity. This was particularly true for what they termed "endogenous attention." Endogenous attention was defined as the cognitive function that selects the relevant pieces of sensory information to achieve goals, and it is known to also be controlled by dorsal fronto-parietal brain areas.

Latent States, Functional Connectivity, and Attention

To better understand this idea that certain networks underpin most forms of human cognitive activity, inclusive of attention, we will briefly look at the research on functional connectivity and latent states. Many of these concepts were generated in computational neuroscience but have their analogs in human brain research, at least in functional connectivity studies.

Historically, functional connectivity (FC) in the human brain was studied by focusing on the brain's resting state. The resting state of the human brain happens when a person is awake and alert but not actively engaged in a task that requires attention or a specific goal. While hypothetically inactive, the brain, in fact, demonstrates an astonishing volume of organized, spontaneous activity. This activity is characterized by a default mode of brain function, where a network of brain regions shows high levels of activity predominantly focused on the resting state, where ongoing dynamics are thought to reflect the brain's intrinsic network architecture, which is thought to be broadly relevant because it persists across brain states (i.e., is state-general). One of these brain states is attention (Visintin et al., 2015).

Recent research has indicated that analyzing brain activity during the resting state may not be the best way to assess FC (McCormick et al., 2022). They proposed that latent FC, which reflects shared connectivity patterns across many brain states, better captures state-general intrinsic FC relative to measures derived from resting state alone. A latent space is a construct that represents a group of neurons with a lower-dimensional space that captures the essential features of a specific, related group of input data, or stimuli. In simpler terms, it is a compressed representation of the original data where each dimension corresponds to a specific feature or characteristic. In human brain research, dimension space refers to how neurons are connected in networks. The number of neurons in a network determines its dimension. Lower dimensionality refers to smaller groups of neurons (Tyukin et al., 2018).

They found that compared with resting-state connectivity, latent FC improves generalization to held-out brain states, better explaining patterns of connectivity and task-evoked activation. Their results indicated that FC patterns shared across many brain states, rather than just resting states, better reflect state-general connectivity. This finding supported the idea of "intrinsic" brain network architecture as a set of connectivity properties persistent across brain states.

Understanding both concepts, the latent state and dimensionality, is one way to look at understanding what is a central goal in cognitive neuroscience research. Specifically, it is a way of understanding how cognition, inclusive of attention, emanates from the dynamic interplay of neural systems. To understand how interactions occur at the level of large-scale functional systems, studies have characterized neural dynamics as a trajectory in a latent state space where each dimension corresponds to the activity of a functional network (Song et al., 2023). This line of research has yielded two central hypotheses. First, neural dynamics operate on a low-dimensional manifold. A neural manifold is a low-dimensional subspace within the higher-dimensional space of neural activity (Mitchell-Hegg et al., 2023). In other words, the dynamics of neural operation can be represented by a small number of independent latent components due to covariation of neural activity within a system. Second, neural activity does not just continuously flow along a singular network; instead, it systematically transitions between recurring latent "states," or hidden clusters, within the state space. Essentially, a finite subset of neural networks underpins the human brain's response to a number of differing tasks. The default attention network is one of these. That certainly makes sense, as, before you attempt to solve a task, you must first pay attention to it. For example, Shine et al. (2019) demonstrated that neural activity traverses a common low-dimensional manifold across seven discrete cognitive tasks. The dynamics within this common manifold were aligned to exogenous task blocks and related to individual differences in cognitive traits.

Research based on latent state modeling and attention has found that the latent states involving attention commonly occurred in many different task types, but their functional roles differed depending on context (Song et al., 2023). They found that the Default Mode Network (DMN) state accompanied successful, stable performance on monotonous tasks, whereas the Dorsal Attention Network (DAN) state characterized suboptimal performance on those same tasks. On the other hand, the DMN state indicated low attentional engagement to narratives, while the DAN state indicated high attentional engagement during documentary watching. In other words, they found that neural state dynamics are modulated by ongoing cognitive and attentional states.

There are emerging clinical implications of dimensionality, neural states, and in some instances, attention. For example, there is research indicating the behavioral effects of a stroke are predictable through low-dimensionality modeling (Corbetta et al., 2018). Their research on stroke, noting that most stroke damage impacts white matter connectivity, led them to hypothesize that alterations in resting FC secondary to focal brain injuries decrease the variability of neural states that the brain can explore both at rest and during tasks. This results in decreased ability to process information effectively. As pertains to treatment, they go on to state that "If an injured brain is unable to generate variable neural states, and if this low variability is expressed in a low-dimensional pattern of behavior and FC, then one goal of rehabilitation, including more recent neuro-stimulation approaches using Transcranial Magnetic Stimulation and Transcranial Direct Current Stimulation, might be able to improve or normalize this variability and restore functional capacity.

Research in dimensionality-related factors in learning disabilities suggests similar kinds of overlapping features of various disorders. The research of Child, Cirino, & Fuchs (2019) supported the conclusion that working memory and phonological awareness were overlapped among reading, math, and attention disorders. They also found that numerosity, the perception of the volume of a number, is related to the comorbidity between math and attention.

Again, it should come as no surprise to recognize that attention underpins task performance. It does begin to generate questions, as yet unanswered, as to the relationship of attention to the deficits of attentional performance in disorders of mental health. While it is certainly possible that an individual's default mode is not operating efficiently, it is also possible that it is operating correctly but being used in the wrong circumstance or being misapplied in relation to goal selection. It is entirely possible that there is nothing wrong with attention at all, but its overall optimal functioning is suboptimal due to the interaction with other networks (i.e., the reward recognition network, the salience network).

The Development of Brain Network Hubs and Attention (See Figs. 5.4 and 5.5)

In order to understand the data emerging from brain mapping research, it would be very nice if we had a model that would suggest how these fundamental transitional processes become integrated into the multiplicity of tasks in which they appear

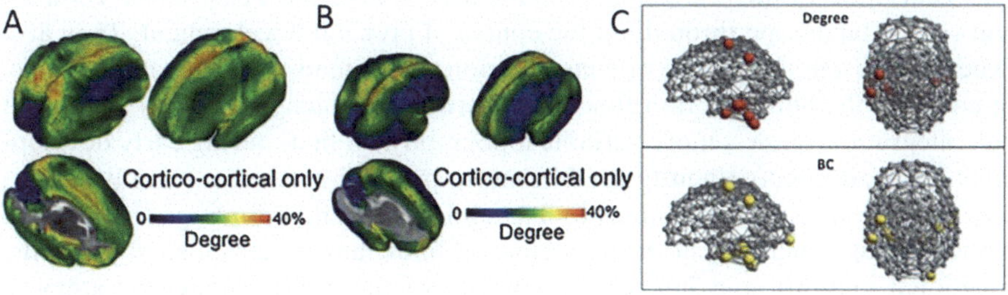

Fig. 5.4 Hubs in prenatal and neonate connectomes. (**a**) Degree of cortical regions in the structural connectivity network of neonates at 40 weeks GA. (**b**) Degree of cortical regions in the structural connectivity network of neonates at 30 weeks GA. These figures show that hubs (red/yellow areas) are present by 30 weeks GA. Several of these hub areas, including superior frontal, superior parietal, and occipital regions, are also hubs in adults. (**c**) Degree and betweenness centrality (BC) hubs in a fetal functional connectivity network (degree hubs are in red and betweenness hubs are in yellow). Functional hubs are primarily localized to primary regions in the prenatal/infant period. In contrast, functional connectivity hubs in adults are located in association areas. Both measures show a high level of consistency in the regions identified as hubs (Oldham et al., 2019). (Panels **a** and **b** adapted from Ball et al. (2014) and panel **c** from van den Heuvel et al. (2018) the web version of this article)

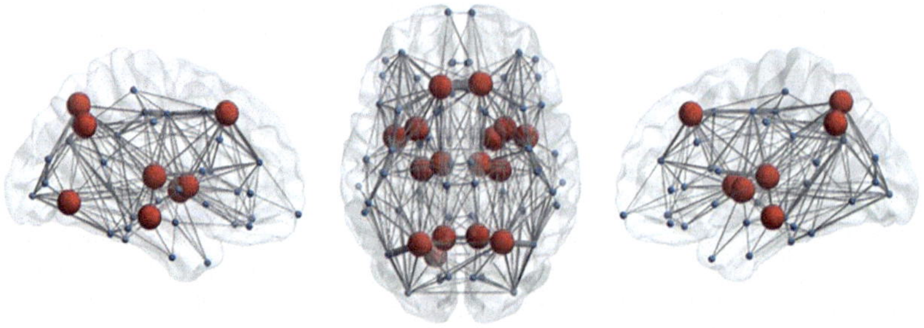

Fig. 5.5 Structural hubs of the adult human brain. (**a**) Spatial distribution of brain network hubs. Here, hubs (red) have been defined as nodes scoring ≥1 standard deviation above the mean on at least one centrality measure (degree, nodal efficiency, or betweenness). Hubs are in the bilateral putamen, thalamus, superior parietal, superior frontal, precuneus, hippocampus, insula, right pallidum, and left lingual gyrus

involved. Fortunately, there is modeling that provides a guideline to understand how this might occur. That modeling is potentially provided by reviewing the data on neural network hubs and their integration into larger neural networks. This model posits that some brain regions have a central role in supporting integrated brain function. This makes them central to both the fully operational brain and results in them being termed network hubs. Given the functional importance of hubs, it is logical to understand when and how they emerge during development and ponder how they shape the function of the maturing brain.

Taken as a whole, the available research points to the fact that structural (physical) hubs of the brain emerge in the prenatal period. These show a consistent spatial landscape throughout the course of prenatal development. They also experience a lengthy period of consolidation that extends into late adolescence. In contrast, the hubs of brain functional networks, which develop as a result of hub integration, show a more variable topography. At first, during early development, they are predominantly established in primary cortical areas. Over time, they migrate to association areas by late childhood. These findings suggest that while the basic anatomical infrastructure of hubs may be established early, the functional viability and integrative capacity of these areas undergo extensive postnatal maturation (Oldham & Fornito, 2019). The factors that impact the development of the function networks are numerous and include genetic (Rolls & Stringer, 2000), epigenetic (Ziller et al., 2015) and life course events (Pharisi & Nolfi, 2001).

From inspection of Fig. 5.1 and 5.2, we can see how these hubs go from relatively isolated centers of activities in infancy to highly integrated functional systems in adults. More importantly for this discussion, we can see that the same core/primary/foundational hubs are integrated into a number of adult systems. A dysfunction in one of these primary hubs will have wide-ranging consequences for later functioning across a number of systems. For example, there is increasing evidence that altered hub topology is identifiable in, at least some disorders, of mental health. These include disorders such as autism, schizophrenia (van den Heuvel et al., 2018) and some instances of intellectual disability (Nelson & Bender, 2021). In addition, there is evidence that disordered hub topography associated with preterm delivery (31 weeks) persists into the neonatal period and is a likely cause of a variety of neurodevelopmental disorders associated with preterm delivery (Rogers et al., 2018). Infants born prematurely, particularly those delivered at the earliest gestational ages, commonly demonstrate increased rates of impairment across multiple neurodevelopmental domains. Preterm birth is a leading risk factor for cerebral palsy, is associated with executive function deficits, increases risk for impaired receptive and expressive language skills, and is linked with higher rates of co-occurring attention deficit hyperactivity disorder, anxiety, and autism spectrum disorders.

Hierarchy of Networks in the Human Neonate

The third trimester is an important period of time for the development of the human connectome. Massive development of cortico-cortical axonal pathways before birth provides a structural basis for the establishment of the adult-like brain network. During that third trimester large-scale neural circuits are formed in the human brain, resulting in a highly efficient and segregated connectome at birth (Zhao et al., 2019). Using constructed structural connectivity matrices and graph-theory-based analyses, they found specific characteristics of network development. Specifically, they found that rapid increases of nodal efficiency were largely centered in the brain hubs distributed in primary sensorimotor regions, superior-middle frontal, and precuneus regions. Their data also indicated that higher rates of nodal edge strength increases were located in the rich-club (dense and highly connected nodes) and within-module connections as compared to other connections. The edge strength (the number of nodes connected to the edge of a node) of short-range connections increased faster than that of long-range connections. They concluded that the above findings supported the idea of early network segregation and differentiated emergence of brain functions. Other research suggested some additional characteristics of third trimester/early neonate network development (Nazari & Salehi, 2023). As pertains to neural networks, their results showed that global efficiency, local efficiency, clustering, and small-worldness increased with age, while modularity and characteristic path length decreased with age.

These characteristics are important to remember as we see that these largely segregated areas at birth commence a lifelong process of integration once the child is born. This integration is achieved by long-range and short-range white matter connections between these primary hubs (nodes). Based on a number of factors, including epigenetics and early environmental influences, these networks develop somewhat specific to each neonate. The pattern of connectivity varies (to a degree) from person to person. This process results in these early/core/fundamental, primary nodes becoming part of many of the routines and subroutines that develop over time. Dysfunction in one of these primary nodes will result in an ever-widening pattern of dysfunction in the routines and subroutines that are developed with their inclusion. Fundamental attention is one set of skills that the neonate possesses, and the visual system that supports if function is one of these primary hubs.

The neuronal activities associated with the circuit formation undergo substantial remodeling after birth. Regionally differentiated developments of neuronal connections result in the heterogeneous emergence of brain functions (Zhao et al., 2019). Primary sensorimotor functions generally emerge earlier than higher-order cognitive functions. Over the course of the first year of life, the neonate brain, originally a largely structural but functionally immature hub network, slowly becomes differentiated into an increasing number of functional modules characterized by the strengthened intra- and inter-modular connectivities. In this process, the original hubs link with other hubs (white matter connectivity), becoming connector hubs which emerge and increase within the original hub; differentiation and specialization also continue. Provincial hubs (the high-centrality regions connecting regions in the same module) diminish. Different hubs have distinct developmental trajectories of the intra- and inter-modular connections, indicating different types of role transitions in the network, such as non-hubs to hubs or provincial hubs to connector hubs, etc. In sum, the research indicates that functional segregation and integration are both increased in the first year of postnatal life. Module reorganization and hub transition lead to more efficient brain networks that are characterized by increasingly segregated modular structure and more connector hubs (Wen et al., 2019).

Research (Molloy, 2022) has resulted in modeling a hierarchy of specific networks with a sequenced emergence pattern in the human neonate. The first network to differentiate itself from the rest of the brain is the sensorimotor network, which includes both the visual and somatomotor cortex. Following this, the visual cortex (both early and late visual cortex) differentiates from the sensorimotor network, followed by further divisions of the visual cortex into a low and high-level visual cortex. As pertains to our discussion of attention, the high-level visual network clearly displays a split into dorsal and ventral streams. The accumulated research data suggests that perhaps the connectivity of these regions is already set up at birth and established to perform these high-level functions (Cabral et al., 2019). The visual and high-level visual clusters then further differentiate into anterior/posterior and medial/lateral divisions. The point here is that the networks responsible for basic attention emerge early on in the process of development and are available for use by the neonate. This, of course, makes sense. The primary areas are integrated with developing, more complex networks as they develop.

All Disrupted Attention Is Not Reflective of Attention Deficit Disorder

We have seen that the neural networks that support attention are present at birth and, through a lifelong process of ongoing integration, are involved in almost every higher-order cognitive activity. We have provided a model that describes how an early dysfunction in one of the components of these primary areas will result in later developing routines and subroutines being dysfunctional.

Simply stated, there will be dysfunctional aspects of attention in many forms of disorders of mental health. These attention difficulties will not always result in behaviors that meet what should be the highly specific behavioral characteristics for a diagnosis of attention deficit disorder under the current, behaviorally based DSM-5nosology. Not all difficulties with attention represent attention deficit disorder. We believe that they deserve empirical study in their own right. Doing so would likely result in new, effective, and targeted treatments for a variety of mental health conditions where disrupted attention is an important clinical feature. Others have highlighted a similar concern (Yiend, 2010) as expressed by the following questions: "What has emotion got to do with attention? Why does the interface between these two warrant additional empirical and theoretical consideration? The answer lies in the many interactions that can occur between emotion and attention and that would otherwise be overlooked by either field in isolation. These interactions are important to basic research. They must be reconciled with existing theories of emotion and attention and may also trigger new theoretical insights. Equally important, research at this interface can inform both the understanding and treatment of various psychopathologies."

References

Ball, G., et al. (2014). Rich-club organization of the newborn human brain. *Proceedings of the National Academy of Sciences of the United States of America, 111*, 7456–7461. https://doi.org/10.1073/pnas.1324118111

Berger, A., et al. (2021). How is the norepinephrine system involved in the antiepileptic effects of vagus nerve stimulation? *Frontiers in Neuroscience, 15*. https://doi.org/10.3389/fnins.2021.790943

Cabral, L., Zubiaurre, L., Wild, C., Linke, A., & Cusack, R. (2019). Category-selective visual regions have distinctive signatures of connectivity in neonates. *BioRxiv*, 675421. https://doi.org/10.1101/675421

Child, A. E., Cirino, P. T., Fletcher, J. M., Willcutt, E. G., & Fuchs, L. S. (2019). A cognitive dimensional approach to understanding shared and unique contributions to reading, math, and attention skills. *Journal of Learning Disabilities, 52*(1), 15–30.

Churchland, M. M., & Shenoy, K. V. (2007). Temporal complexity and heterogeneity of single-neuron activity in premotor and motor cortex. *Journal of Neurophysiology, 97*, 4235–4257.

Corbetta, M., & Shulman, G. (2002). Control of goal-directed and stimulus-driven attention in the brain. *Nature Reviews Neuroscience, 3*, 201–215.

Corbetta, M., & Shulman, G. L. (2011). Spatial neglect and attention networks. *Annual Review of Neuroscience, 34*(1), 569–599.

Corbetta, M., Siegel, J. S., & Shulman, G. L. (2018). On the low dimensionality of behavioral deficits and alterations of brain network connectivity after focal injury. *Cortex, 107*, 229–237.

Dosenbach, N., Fair, D., Cohen, A., Schlaggar, B., & Petersen, S. (2008). A dual-networks architecture of top-down control. *Trends in Cognitive Neuroscience, 12*, 99–105. https://doi.org/10.1016/j.tics.2008.01.001

Duménieu, M., Marquèze-Pouey, B., Russier, M., & Debanne, D. (2021). Mechanisms of plasticity in subcortical visual areas. *Cells, 10*(11), 3162. https://doi.org/10.3390/cells10113162

Evarts, E. V. (1868). Relation of pyramidal tract activity to force exerted during voluntary movement. *Journal of Neurophysiology, 31*, 14–27.

Fortunato, C., Bennasar-Vázquez, J., Park, J., Chang, J., Miller, L., Dudman, J., et al. (2024). Nonlinear manifolds underlie neural population activity during behaviour. *bioRxiv [Preprint]*. 2024 Apr 25:2023.07.18.549575. https://doi.org/10.1101/2023.07.18.549575. PM.

Fuster, J. (1997). Network memory. *Trends in Neuroscience, 20*(10), 451–459. https://doi.org/10.1016/S0166-2236(97)01128-4

Gallego, J., Perich, M., Naufel, S., Ethier, C., Solla, S., & Miller, E. (2018). Cortical population activity within a preserved neural manifold underlies multiple motor behaviors. *Nature Communications, 9*, 4233. https://doi.org/10.1038/s41467-018-06560-z

Gerstner, W., Kistler, W., Naud, R., & Paninsk, L. (2014). Neuronal populations. In W. Gerstner, W. Kistler, R. Naud, & L. Paninsk (Eds.), *Neuronal dynamics from single neurons to networks and models of cognition* (pp. 291–322). Cambridge University Press.

Holland, N., Robbins, T., & Rowe, J. (2021). The role of noradrenaline in cognition and cognitive disorders. *Brain, 144*, 2243–2256. https://doi.org/10.1093/brain/awab111

Humphrey, D. R., & Thompson, W. D. (1970). Predicting measures of motor performance from multiple cortical spike trains. *Science, 170*, 758–762.

Jensen, K. T., Kadmon Harpaz, N., Dhawale, A. K., Wolff, S. B., & Ölveczky, B. P. (2022). Long-term stability of single neuron activity in the motor system. *Nature Neuroscience, 25*(12), 1664–1674.

Markett, S., Nothdurfter, D., Fosca, A., & Reuter, M. (2022). Attention networks and the intrinsic network structure of the human brain. *Human Brain Mapping, 43*, 1431–1448. https://doi.org/10.1002/hbm.25734

McCormick, E., Arnemann, K., Ito, T., Hanson, S., & Cole, M. (2022). Latent functional connectivity underlying multiple brain states. *Network Neuroscience, 6*(2), 570–590. https://doi.org/10.1162/netn_a_00234

Meyer, K., & Damasio, A. (2009). Convergence and divergence in a architecture for recognition and memory. *Trends in Neurosciences, 32*(7), 376–382. https://doi.org/10.1016/j.tins.2009.04.002

Mitchell-Hegg, R., Prado, S., Gava, G., Go, M., & Schultz, S. (2023). Neural manifold analysis of brain circuit dynamics in health and disease. *Journal of Computational Neuroscience, 51*(1), 1–21. https://doi.org/10.1007/s10827-022-00839-3

Molloy, M. S. (2022). Individual variability in functional organization of the neonatal brain. *NeuroImage, 253*, 119101. https://doi.org/10.1016/j.neuroimage.2022.119101

Nazari, R., & Salehi, M. (2023). Early development of the functional brain network in newborns. *Brain Structure & Function, 228*, 1725–1739. https://doi.org/10.1007/s00429-023-02681-4

Nelson, A., & Bender, K. (2021). Dendritic integration dysfunction in neurodevelopmental disorders. *Developmental Neuroscience, 43*(3–4), 201–221. https://doi.org/10.1159/000516657

Oldham, S., & Fornito, A. (2019). The development of brain network hubs. *Developmental Cognitive Neuroscience, 36*, 100607.

Oldham, S., Fulcher, B., Parkes, L., Arnatkevic̆iūtė, A., Suo, C., & Fornito, A. (2019). Consistency and differences between centrality measures across distinct classes of networks. *PLoS One, 14*(7), e0220061. https://doi.org/10.1371/journal.pone.0220061

Petersen, S., & Posner, M. (2012). The attention system of the human brain: 20 years after. *Annual Review of Neuroscience, 35*, 73–89. https://doi.org/10.1146/annurev-neuro-062111-150525. PMID: 22524787; PMCID: PMC3413263.

Petersen, S., Fox, P., Miezin, F., & Raichle, M. (1988). *Modulation of cortical visual responses by direction of spatial attention measured by PET* (p. 22). Association for Research in Vision Ophthalology.

Pharisi, D., & Nolfi, F. (2001). *Development in neural networks*. Retrieved from MIT Press Direct. https://direct.mit.edu/books/edited-volume/2039/chapter-abstract/55299/Development-in-Neural-Network

Posner, M., & Petersen, S. (1990). The attention system of the human brain. *Annual Review of Neuroscience, 13*(1), 25–42. https://doi.org/10.1146/annurev.ne.13.030190.000325

Rogers, T., Goldstein, N. J., & Fox, C. R. (2018). Social mobilization. *Annual Review of Psychology, 69*(1), 357–381.

Rolls, E. T., & Stringer, S. M. (2000). On the design of neural networks in the brain by genetic evolution. *Progress in Neurobiology, 61*(6), 557–579. https://doi.org/10.1016/S0301-0082(99)00066-0

Saboory, E., Ghasemi, M., & Mehranfard, D. (2020). Norepinephrine, neurodevelopment and behavior. *Neurochemistry International, 135*, 104706. https://doi.org/10.1016/j.neuint.2020.104706

Sani, I., Stemmann, H., Caron, B., et al. (2021). The human endogenous attentional control network includes a ventro-temporal cortical node. *Nature Communications, 12*, 360. https://doi.org/10.1038/s41467-020-20583-5

Scott, S. H., & Kalaska, J. F. (1997). Reaching movements with similar hand paths but different arm orientations. I. Activity of individual cells in motor cortex. *Journal of Neurophysiology, 77*, 826–852.

Shine, J. M., Hearne, L. J., Breakspear, M., Hwang, K., Müller, E. J., Sporns, O., ... & Cocchi, L. (2019). The lowdimensional neural architecture of cognitive complexity is related to activity in medial thalamic nuclei. *Neuron, 104*(5), 849–855.

Singer, W. (1993). Synchronization of cortical activity and its putative. *Annual Review of Physiology, 55*, 349–374.

Song, H., Shim, W., & Rosenebrg, M. (2023). Large-scale neural dynamics in a shared low-dimensional state space reflect cognitive and attentional dynamics. *Neuroscience*. https://doi.org/10.7554/eLife.85487

Stam, C. (2024). Hub overload and failure as a final common pathway in neurological brain network disorders. *Network Neuroscience, 8*(1), 1–23. https://doi.org/10.1162/netn_a_00339

Tyukin, I., Gorban, A., Calvo, C., Makarova, J., & Makarov, V. (2018). High-dimensional brain: A tool for encoding and rapid learning of memories by single neurons. *Bulletin of Mathematical Biology, 81*(11), 4856–4888. https://doi.org/10.1007/s11538-018-0415-5

van den Heuvel, M., & Sporns, O. (2013). Network hubs in the human brain. *Trends in Cognitive Sciences, 17*(12), 683–696. https://doi.org/10.1016/j.tics.2013.09.012

van den Heuvel, M., et al. (2018). Hubs in the human fetal brain network. *Developmental Cognitive Neuroscience*. https://doi.org/10.1016/j.dcn.2018.02.001

Visintin, E., De Panfilis, C., Antonucci, C., Capecci, C., Marchesi, C., & Sambataro, F. (2015). Parsing the intrinsic networks underlying attention: A resting state study. *Behavioural Brain Research, 278*, 315–322. https://doi.org/10.1016/j.bbr.2014.10.002

Weber, P., Wagner, W., & Freitag, S. (2023). Physically enhanced training for modeling rate-independent plasticity with feedforward neural networks. *Computational Mechanics, 72*(4), 827–857.

Wen, X., et al. (2019). First-year development of modules and hubs in infant brain functional networks. *NeuroImage, 185*, 222–235. https://doi.org/10.1016/j.neuroimage.2018.10.019

Xuan, B., Mackie, M., Spagna, A., Wu, T., Tian, Y., Hof, P. R., & Fan, J. (2016). The activation of interactive attentional networks. *NeuroImage*, 308–319. https://doi.org/10.1016/j.neuroimage.2016.01.017

Yiend, J. (2010). The effects of emotion on attention: A review of attentional processing of emotional information. In J. de Houwer & D. Hermans (Eds.), *Cognition & emotion. Current research and theories*. Psychology Press.

Zhao, T., Mishra, V., et al. (2019). Structural network maturation of the preterm human brain. *NeuroImage, 185*, 699–710. https://doi.org/10.1016/j.neuroimage.2018.06.047

Ziller, M. J., Edri, R., Yaffe, Y., Donaghey, J., Pop, R., Mallard, W., ... & Meissner, A. (2015). Dissecting neural differentiation regulatory networks through epigenetic footprinting. *Nature, 518*(7539), 355–359.

Zohary, E., Shadlen, M., & Newsome, W. (1994). Correlated neuronal discharge rate and its implications for psychophysical performance. *Nature, 370*, 140–143.

Chapter 6
Top-Down, Bottom-Up Attentional Processing. The Traditional View

The human brain is a marvelously complex processing machine. It contains about 100 billion neurons, and most research suggests a firing rate for each neuron of 0.29 per second (Le-Cunff, 2022). Each neuron is hypothesized to connect to about 7000 other neurons. This means that each time a specific neuron fires a signal, 7000 other neurons receive the information. If you multiply these three numbers, you get 200 million bits of information transmitted every second inside your brain. Despite this massive computing power, the human brain is actually limited in its capacity to process all sensory stimuli present in the environment surrounding the individual at any point in time. It has to narrow the stimulus field and identify things that are important and things that are not. This, at the most basic level, ensures the survival of the individual. This process of selecting a subset of the input, and ignoring the rest, is referred to as attention. The brain relies on the cognitive process of attention to focus neural resources according to the exigencies of the moment.

Traditionally, attention was segmented into two distinct groups of functions. The first, bottom-up attention, refers to attentional allocation to sensory stimuli external to the individual, which are salient because of their inherent properties relative to the background. Think of a person wearing a red dress surrounded by all black dresses at a dinner party. The second, top-down attention, refers to internal management of attention based on prior knowledge, willful plans, and current goals (Katsuki & Constantinidis, 2013). Consider that this person is the one you came to meet, so your attention is focused on how to meet up with them in particular. Although dissociable neural network processes mediate the allocation of attention based on bottom-up and top-down factors, a common neural apparatus, the frontoparietal network, necessary in cognitive control, is common in both types of attentional processes. Top-down and bottom-up modeling is frequently utilized in the study of visual attention, wherein attention can be focused volitionally by top-down signals determined by the current task (e.g., looking for something) and automatically by

T. Wasserman, L. D. Wasserman, *Disorders of Attention*, Neural Network Model: Applications and Implications, https://doi.org/10.1007/978-3-031-78732-4_6

bottom-up signals associated with unexpected, novel, or salient stimuli, such as a red dress popping out (Anderson et al., 2018).

Again, top-down attention refers to the voluntary allocation of attention to certain features, objects, or regions in space (Pinto et al., 2013). Attention affects the mean neuronal firing rate as well as its variability and correlation across neurons. As we have seen, although distinct neural network processes facilitate the direction of attention based on bottom-up and top-down factors, there is a common neural apparatus, the frontoparietal network, essential in both types of attentional processes (Katsuki & Constantinidis, 2013). Therefore, while these two attentional processes are often described individually and they are hypothesized to involve distinct neural mechanisms and anatomic substrates, it is important to remember that in everyday experience, bottom-up and top-down factors constantly influence each other to orient attention. For example, there is evidence that difficulties with both top-down and bottom-up processing are associated with attention deficit hyperactivity disorder (Schneidt et al., 2018), a disorder whose presentation includes difficulties with cognitive control.

It is important to note that, for the most part, newborns use bottom-up attention. For newborns, novelty is crucial for the development of attentional capacity early on, and bottom-up processing is the basis for all future attentional development. This is part of what we call the sensory motor stage and is the beginning of integration across arousal, focused attention, emotional regulation, reward history, and effort. Once there is a sensory-motor base, attention becomes more top-down-oriented. We will be focusing on top-down processing for the majority of this book. Given our network orientation and emphasis on the reward recognition network and its role in attention, that makes sense.

Similarities and Differences

There is a similarity between top-down and bottom-up deployments of attention. First and foremost is that, although the reason for attentional deployment is different, the effects of the two are largely the same. In both cases, the attended objects receive preferential processing, and, in both cases, this leads to an increased neural response, which has functional consequences, such as developing better memory storage. There are also important differences between both types of attention. Top-down attention is also referred to as endogenous or sustained attention, and bottom-up attention is commonly typified as exogenous or transient attention (Carrasco, 2011). Top-down attention is labeled endogenous because, unlike bottom-up attention, which is considered automatic/involuntary, it is under clear voluntary control. Importantly, top-down attention is more sustained, as individuals typically direct their top-down attention at stimuli for sustained periods of time, whereas bottom-up attention is transiently captured. Finally, top-down attention seems to take longer to deploy than bottom-up attention.

In reality, segmenting these two processes is for educational purposes, as, almost from the beginning, human infants form rudimentary templates based on pattern or facial recognition and impose top-down deliberations on bottom-up information. These systems become increasingly intertwined as a person develops, as the following describes: "How much are we in control of our decisions? The current study explores how goal-relevant (top-down) and perceptual features (bottom-up) govern the distribution of attention in risky choice. Findings from participants' choices, eye movements, and computational modeling analysis show that while salient perceptual features can influence the acquisition of information, their influence on choice is limited to situations in which those features are tied to potential outcomes of the decision. When such features are not tied to outcomes, their influence is suppressed by top-down control. These results suggest, contrary to popular accounts, that perceptual salience can induce bottom-up effects of overt selection, but the perceived value of information is the crucial arbiter of intentional control over choice" (Vanunu et al., 2021).

Bottom-Up Attention Processing

As might be expected from a process present at birth in the human infant, bottom-up attention is, from the beginning, largely but not exclusively a visually based process. Visual attention is a cognitive process that enables human infants to rapidly select highly relevant information from a scene. Later in development, this information is then processed more finely by high-level, top-down-based visual processes such as scene understanding and object recognition. The notion of relevance is determined by two factors. The first one, often referred to as bottom-up visual saliency, is a task-independent component based on only low-level and image-based outliers and easily noted differences (Borji, 2012). Visual attention starts with basic visual processing along the visual cortical pathways. From the primary visual cortex, feed-forward signals ascend to multiple cortical areas and branch into two major visual pathways: a ventral pathway dealing with object- and feature-based visual processes, and a dorsal pathway dealing with spatial- and movement-related visual processes. Areas that are part of the ventral pathway include the inferior temporal (IT) cortex, which projects to the ventral part of the lateral prefrontal cortex (PFC). The dorsal pathway comprises the middle temporal (MT) and medial superior temporal (MST) areas, as well as areas in the posterior parietal cortex (PPC), which project to the dorsolateral part of the PFC (dlPFC).

The two cortical pathways, the ventral and the dorsal pathways, are organized across a cortical hierarchy in which receptive fields grow bigger and functional properties become more complex as inputs ascend from the earlier stages to the later stages of the pathways. Bottom-up information that can guide attention propagates thus from the visual cortex to the PFC (Katsuki & Constantinidis, 2013).

Putting the importance of the neural networks into perspective, it is important to note that the PFC is involved in the temporal organization of behavior, supporting cognitive functions necessary to organize behavior in time and across contexts. Looked at from this functional perspective, one can see the importance of this pathway as it impacts cognitive control across developmental stages. For the PFC region, this is an extended process. It has been posited that the period from early childhood to kindergarten is when the PFC forms the basic neural circuitry that will underlie higher cognitive functions later on (Tsujimoto, 2008). Early exposure to life factors and early experiences set up the PFC neural developmental trajectories and can have lifelong consequences on behavioral regulation, including the regulation of attention. Remembering that experiences can be positive or adverse, and the development prolonged, the effect on the regulatory control of the PFC region can develop in a neurotypical fashion or be susceptible to factors which lead to atypical or abnormal development and functioning (Kolb, 2012).

Visual Competition and Salience

From a sensory perspective, objects that differentiate themselves stand out and are selected for attention. This selection process can be described as competitive based on a number of factors. As a result of competition, the most salient stimulus is represented predominantly at all stages of the visual system and leads to bottom-up orienting of attention. A related construct, saliency, refers to all types of feature information (e.g., shape, color, and orientation) in the visual field that are pooled, and attention is allocated to the location with the highest activity/novelty. These saliency factors can be represented as saliency maps. There are multiple models that attempt to represent saliency variables. Some are computational, some are based on eye tracking data, and others blended. Suffice it to say that there is yet no consensus on how this process works.

It remains unclear then as to the contribution to attention these bottom-up processes play in later development. There is mixed evidence to suggest that these perceptual features can play a role under very specific circumstances to, at least, attract initial attention. Research has shown that initial (bottom-up) attraction to elements of a display that are visually salient, such as numbers appearing in a larger font size, objects with distinctive physical features (different in size from other items in an array), or objects that were more "intense" than those of other items (e.g., more white pixels, subtended a greater visual angle, deeper color, or a different color from other items in the array), can capture initial attention but are then modifiable with top-down information (Vanunu et al., 2021).

There is a caveat to the above in reference to the apparently rather clear relationship of top-down to bottom-up processing. There are situations where top-down processes cannot exert control over bottom-up information flow. These issues related to the limited capacity of the top-down executive suppression capabilities. Recent research demonstrated that the relative salience of items in the display is a crucial factor in attentional control. In displays with a few heterogeneous items, top-down suppression is possible. However, in larger displays in which both target and distractor singletons are salient, no top-down suppression is observed. Specifically, in this study, a paradigm showed that while searching for a specific shape, observers suppressed the location of the irrelevant color singleton. It also showed that this suppression, due to top-down suggestion to ignore color, occurs only for search arrays with four elements. For larger array sizes, when both target and distractor singleton are salient, there is no evidence for suppression; instead, there is clear evidence that the salient distractor captured attention (Wang & Theeuwes, 2020).

Bottom-Up Processing, Attention Disorder, and Executive Function

The above exception suggests a model of attentional deficit that is consistent with historical modes of the disorder. That is, bottom-up environmental information overwhelms the central executive and captures attention. This, of course, can happen in a number of ways. The first is that there is a plethora of environmental information that exceeds the capacity of the central executive to manage the information flow (bottom-up). The second, a more likely explanation, is that the central executive in that person is inefficient. These views of the development of executive function (EF) suggest that EF is an early manifested aspect of the developing infant. There is another perspective, one that we support, that EF initially develops as a way to control automated and foundational motor behavior (Koziol & Lutz, 2013). They proposed that the functional architecture of the brain that subserves EF evolved to meet the needs of interactive behavior and that higher cognition develops to control the motor system. They hypothesized that traditional models of cognition are incomplete characterizations of EF and that procedural learning and "automatic" behaviors are the most basic, bottom-up functions that support all EF. In sum, their model outlined a bottom-up process wherein motor development in childhood models how all knowledge is grounded in sensorimotor interaction and how interactive behavior generates both procedural and declarative knowledge. This process generates and stimulates the development of EF.

Although we are suggesting a direct causal link, there is considerable research supporting the idea that ADHD is, at least, associated with poor executive function (Biederman et al., 2004; Scahachar, 2023). Biederman et al. note that a literature review of 18 studies concluded that children with ADHD consistently exhibit worse performance on certain cognitive and EF measures. One aspect of the executive function deficit involves the failure to inhibit extraneous and likely bottom-up information. In reference to planning, successful execution of a plan largely involves eliminating distracting activities and stimuli. This involves inhibiting these distracting influences and is accomplished through our prefrontal inhibitory centers. The result of the inhibition is to allow an individual time during which they can consider options before reacting. This lack of inhibition is a double problem for people with

ADHD. First, they will be unable to adequately inhibit distractions, inhibit impulsive reactions, or inhibit physically acting upon these stimuli [hyperactivity]. Second, there are longer-term consequences in that individuals with ADHD do not inhibit their behavior long enough for the other executive functions below to develop adequately (Zorcec & Jordanova, 2010). In fact, most studies on the failure of executive management note that the failure results in the lack of inhibition of extraneous information. By extrapolation, we can conjecture that much of that extraneous information is the result of bottom-up processing.

To be clear, we are not proposing that all attention deficit disorders of this type result from bottom-up information overwhelming a weak central executive, but we are suggesting that at least some of them do. This view is supported by research suggesting that while ADHD is associated with significant weaknesses in several key EF domains, moderate effect sizes and the lack of universality of EF deficits among individuals with ADHD suggest that EF weaknesses are neither necessary nor sufficient to cause all cases of ADHD (Willcut et al., 2005). We would be remiss if we didn't point out an intriguing possibility. That is, that early problems with the regulation of attention might be responsible for the problems with executive function demonstrated in the ADHD population. Research has shown that difficulties with set shifting, for example, are often precursors of executive function deficits later on (Mulder et al., 2009).

Whatever the etiologic sequence, it is clear that issues with executive management abound. In addition to inhibitory issues, there are others, including verbal and visuospatial working memory, the ability to shift one's attention from one task to another, sustained attention, delay aversion, decision-making, verbal fluency, and planning. There are also issues with the executive management system that cause poor regulation of top-down processes that occur that have nothing to do with bottom-up processing (see the Chap. 8, for example). There are several of these, which will be covered separately. In summation at the moment, we have looked at how an organism begins with a heavy bottom-up emphasis, which begins to shift rapidly as the organism matures. This process is prolonged and intertwined with other cortical pathways, as the intention of development is healthy integration. Damage to this process can result in executive function deficits of the kind often seen in ADHD, which is a neurodevelopmental disorder. This type of disruption, however, can be presented across psychopathologies and across age stages.

References

Anderson, P., He, X., Buehler, C., Teney, D., Johnson, M., & Gould, S. (2018). Bottom-up and top-down attention for image captioning and visual question answering. In *CVPR proceedings of the IEEE conference on computer vision and pattern recognition* (pp. 6077–6086). IEEE.

Biederman, J., Monuteaux, M., Doyle, A., Seidman, L., Wilens, T., Ferrero, F., et al. (2004). Impact of executive function deficits and attention-deficit/hyperactivity disorder (ADHD) on academic outcomes in children. *Journal of Consulting and Clinical Psychology, 72*(5), 757–766. https://doi.org/10.1037/0022-006X.72.5.757

Borji, A. (2012). Boosting bottom-up and top-down visual features for saliency estimation. In *IEEE conference on computer vision and pattern recognition* (pp. 438–445). IEEE.

Carrasco, M. (2011). Visual attention: The past 25 years. *Vision Research, 51*(13), 1484–1525.

Katsuki, M., & Constantinidis, C. (2013). Bottom-up and top-down attention: Different processes and overlapping neural systems. *The Neuroscientist, 20*(5), 509–522. https://doi.org/10.1177/1073858413514136

Kolb, B. M. (2012). Experience and the developemnt of the prefrontal cortex. *Proceedings of the National Academy of Science, 109*, 17186–17193.

Koziol, L., & Lutz, J. (2013). From movement to thought: The development of executive function. *Applied Neuropsychology: Child, 2*(2), 104–115. https://doi.org/10.1080/21622965.2013.748386

Le-Cunff, A. (2022). *Is there a limit to the human brain's capacity?* Retrieved from nesslabs. https://nesslabs.com/brain-limitations

Mulder, H., Pitchford, N., Hagger, M., & Marlow, N. (2009). Development of executive function and attention in preterm children: A systematic review. *Developmental Neuropsychology, 34*(4), 393–421. https://doi.org/10.1080/87565640902964524

Pinto, Y., van der Leij, A., Sligte, S., Lamme, V., & Scholte, S. (2013). Bottom-up and top-down attention are independent. *Journal of Vision, 13*, 16. https://doi.org/10.1167/13.3.16

Scahachar, R. (2023). Fifty years of executive control research in attention-deficit/hyperactivity disorder:What we have learned and still need to know. *Neuroscience & Biobehavioral Reviews, 155*, 105461. https://doi.org/10.1016/j.neubiorev.2023.105461

Schneidt, A., Jusyte, A., Rauss, K., & Schönenberg, M. (2018). Distraction by salient stimuli in adults with attention-deficit/hyperactivity disorder: Evidence for the role of task difficulty in bottom-up and top-down processing. *Cortex, 101*, 206–220. https://doi.org/10.1016/j.cortex.2018.01.021

Tsujimoto, S. (2008). The prefrontal cortex: Functional neural development during early childhood. *The Neuroscientist, 14*, 345–358.

Vanunu, Y., Hotaling, J., Le Pelley, M., & Newell, B. (2021). How top-down and bottom-up attention modulate risky choice. *PNAS, 118*(39). https://doi.org/10.1073/pnas.2025646118

Wang, B., & Theeuwes, J. (2020). Salience determines attentional orienting in visual selection. *Journal of Experimental Psychology: Human Perception and Performance, 46*(10), 1051–1057. https://doi.org/10.1037/xhp0000796

Willcut, E., Doyle, A., Nigg, J., Faraone, S., & Pennington, B. (2005). Validity of the executive function theory of attention-deficit/hyperactivity disorder: A meta-analytic review. *Biological Psychiatry, 57*(11), 1336–1346.

Zorcec, T., & Jordanova, N. (2010). ADHD as an executive dysfunction. *Contributions, Section Biological. Medical Science, 31*(2), 171–181. UDK: 159.922.7:616.8-008.6

Chapter 7
Attentional Control

Attentional control (AC) is described as the general ability to control attention in relation to positive or negative information (Derryberry & Reed, 2002). AC comprises focusing and shifting attention. Attention involves three functional processes, which differ neuroanatomically through areas of white matter connectivity and represent distinct neural networks: alerting, orienting, and executive control. Depending on the research, these same processes have been labeled alertness, selectivity, and processing capacity. Posner and Boies (1971) further described them as alertness, selection, and limited processing capacity and stressed that they were separable and dissociable components of attention. The importance here is that the study of attention is attempting to move to better definitions and distinctions of the construct and the underlying processes. As such, individuals can experience difficulty within various points within these attention systems, while others may show strengths and weaknesses in different systems and within systems, depending upon both internal and external variables.

Attention Control Theory

Attention control theory is not a general theory of attentional control but instead is concerned with attentional control in the context of anxiety and cognitive performance. It is an extension of processing efficiency theory (Eysenck & Calvo, 1992). Processing efficiency theory focused on the distinction between effectiveness and efficiency. Effectiveness references the quality of task performance indexed by standard behavioral measures (i.e., response accuracy). In comparison, processing efficiency refers to the relationship between the effectiveness of performance and the effort or resources required or spent in order to achieve the level of task performance. Typically, efficiency decreases as more resources are invested to attain a

T. Wasserman, L. D. Wasserman, *Disorders of Attention*, Neural Network Model: Applications and Implications, https://doi.org/10.1007/978-3-031-78732-4_7

given performance level. As indicated, processing theory also focuses on the relationship of anxiety to attention. It should be stated at the outset that anxiety here references typical anxiety experienced by individuals that do not meet the criteria for anxiety disorder. It is the typical anxiety experienced by individuals when they encounter threats in the environment and the relationship between the perception of the threat and the impact on the effectiveness of their behavior or the impact on the expenditure of resources.

The current authors wish to highlight the potential impact of such processes on both neuropsychological data and clinical implications. Specifically, if there is an impact in neurotypical populations, what, then, can we expect would be the effect in more highly stressed populations or under more highly stressful situations? For example, what are the effects on a person taking a high-stakes exam, such as for college entrance? What is the effect of taking a high-stakes exam on a person suffering from generalized anxiety? And do the neuropsychological constructs we use speak to these differences? Again, is anxiety a unified construct? This brings us to processing efficiency theory.

Processing Efficiency Theory

Processing efficiency theory speaks to the relationship between performance effectiveness and the effort needed to attain said level of performance. Processing efficiency theory rests on two major assumptions. The first is that worry is the component of state anxiety responsible for effects of anxiety on performance effectiveness and efficiency. Worry, defined as self-preoccupation, is characterized by concerns over evaluation and failure and expectations of aversive consequences. Worry is activated in stressful situations and is most likely to occur in individuals high in trait anxiety (Eysenck, 1992). Worry can have two effects. A negative effect involves cognitive interference by preempting the processing and temporary storage capacity of working memory. Worrisome thoughts consume the limited attentional resources of working memory, which results in less available resources for concurrent task processing. The other postulate is one of a positive effect, as worry may enhance motivation to minimize the aversive anxiety state. This is accomplished by promoting increased effort and use of auxiliary processing resources and strategies to promote efficiency. If these resources are unavailable, then performance effectiveness will be impaired.

A second key prediction of processing efficiency theory is that high levels of trait anxiety impair processing efficiency more than performance effectiveness because individuals who experience high levels of anxiety often utilize compensatory strategies or mechanisms to overcome inefficiency caused by non-task processing without concomitant performance improvement. The prediction posits that the main effects of anxiety are on the central executive, a system involved in complex functions including planning and attention. Inherent in the process is the involvement of language and visuospatial processes, particularly the phonological loop and the storing of visual and spatial information. There are then detrimental effects of anxiety also expected on the phonological loop. As a result, adverse effects of anxiety on performance and efficiency should be greater on tasks imposing substantial demands on the processing and storage capacity of working memory. There is an additional burden on the self-regulatory mechanism inhibiting such thoughts and producing auxiliary processing activities. Specifically, processing efficiency theory predicts that trait anxiety impairs three executive functions: the inhibition function (using attentional control to inhibit focus on task-irrelevant stimuli and prepotent but incorrect responses); the shifting function (using attentional control to redirect attention in response to changing demands within and between tasks); and the updating function (updating and monitoring information within working memory) (Berggren & Derakshan, 2013).

Attentional control theory represents a significant development of processing efficiency theory, building on its strengths and addressing some of its limitations. The key assumption that there is an important distinction between processing efficiency and performance effectiveness is also central to attentional control theory. AC theory extends the scope of the processing efficiency theory and is more precise about effects of anxiety on the functioning of the central executive (Derryberry & Reed, 2002; Eysenck et al., 2022).

AC theory posits that impaired efficiency of these executive functions is controlled by interactions between two attentional systems that figure prominently in our discussions. These are a top-down, "goal-directed attentional system influenced by expectation, knowledge, and current goals" and a bottom-up, "stimulus-driven attentional system responding maximally to salient or conspicuous stimuli" (Corbetta & Shulman, 2002).

Neural Network Components of Attention Control

Several functional brain networks have been identified in attention control. Goal-driven, top-down attentional control is known to be subserved by activation in the dorsolateral prefrontal cortex (DLPFC) and a wider network of superior parietal regions known as the frontoparietal network (FPN) (Dosenbach et al., 2008). A second compensatory or "reactive" attentional control system is employed when task-elated errors and conflict are detected (Braver, 2012). The Cingular Opercula Network (CON), also referred to as the salience network (Seeley, 2019), provides a late correction mechanism that maintains and facilitates optimal task performance. CON activity, particularly the dorsal anterior cingulate (dACC), sends signals to the DLPFC to update task demands.

The distinction between orienting, alerting, and executive control is identifiable based upon neural connectivity differences (Posner & Petersen, 1990). The orienting neural network allocates attention to relevant areas of the environment and involves the posterior parietal lobe, the superior colliculus, and the thalamus (Posner & Raichle, 1994). The alerting network comprises the right thalamic, frontal, and parietal regions and the reticular formation (Sturm & Willmes, 2001). Using neuropsychological data, they described alerting as subserved by right hemispheric areas with parieto-fronto-thalamic interactions. Alerting is described as influenced by the cortical distribution of the brain's norepinephrine system (Coull et al., 1996). The executive network is subserved by the anterior areas of the frontal cortex, including the anterior cingulate cortex (Pardo & Janer, 1990) and the lateral prefrontal cortex, which are target areas of the ventral tegmental dopamine system (Benes, 2000).

Elements of the default mode network (DMN), such as the posterior cingulate gyrus (PCC; part of the medial DMN), appear to be strongly associated with worry (Makovac et al., 2020). The PCC is also involved in lapses in attention control (Cieslik et al., 2015).

ACT stresses the differential involvement of these three brain networks (DMN, CON, and FPN) during task performance by individuals with high and low trait anxiety or worry. Neural activity in the FPN is negatively correlated with activity in the DMN. Activity in the FPN increases due to attentional demands; activity in the DMN decreases (or deactivates) over the duration of a task requiring some attentional control. During high load, where compensatory processes are required, activity in CON regions increases to maintain effective task performance by indicating the need for increased attentional control. CON activity is required to compensate for altered FPN activity (Eysenck et al., 2022).

These same networks participate in other attention-related activities, such as the determination of salience. This should not be surprising, as various combinations of networks are recruited for differing processes.

There are white matter connectivity issues associated with anxiety and related attention control. As an example, high anxiety is associated with reduced functional connectivity between ACC and DLPFC (or CON and DLPFC) (Barker et al., 2018). Dysfunctional connectivity occurs in high worriers (Weber-Goericke & Muehlhan, 2019). High worriers had significant activation in brain areas connected to language processing (e.g., middle frontal gyrus; inferior frontal gyrus; anterior insula). There were strong functional connections between these areas and DMN. It is therefore likely that high worriers may have strong connections among brain areas associated with ongoing internally oriented task-irrelevant processing. These strong connections may affect network switching and interactions during task processing and reflect their impaired attentional control and processing inefficiency.

In summary, there is evidence from both behavioral and neuroimaging sources indicating that anxiety affects central executive functioning, particularly impairing efficient functioning of inhibition and shifting processes, with processing efficiency impacted more so than performance effectiveness.

Attention Control and ADHD

High-demanding tasks stress attention control mechanisms responsible for allowing the selection of information from one (focused attention) or multiple (divided attention) sources. These networks react to changes in a given situation (stimulus-driven attention) and allocate effort according to demands (task-positive and task-negative activity) (Salmi et al., 2018). Salmi et al. (2018) found that during auditory-focused attention, there was higher activation in the sensory cortical areas of irrelevant modality and the default mode network (DMN). DMN activity also increased during divided attention in an ADHD group. Adding irrelevant stimulation resulted in enhanced activity in the salience network. Irrelevant distractors that capture attention in a stimulus-driven manner activated dorsal attention networks and the cerebellum. In sum, this research implied that attention control deficits involve the activation of irrelevant sensory modality, problems in regulating the level of attention on demand, and may encumber top-down processing in cases of irrelevant information. Although the authors suggest that the ADHD caused the deficits in attention control, a corollary is also possible in that the issues in attention control might be responsible for the behavioral deficits that lead to the ADHD diagnosis.

There are certainly other attention control mechanisms identified in individuals with ADHD. Schachar et al. (2000) found that deficient inhibitory control distinguishes children with a diagnosis of attention-deficit/hyperactivity (ADHD) disorder, conduct disorder (CD), and comorbid ADHD + CD from normally developing children. Specifically, they found that the ADHD group had significantly impaired inhibitory control compared to (control) NC, CD, and ADHD + CD children. Inhibition was defined as the inability to stop an action and allow an individual to engage in another action. They considered their results as indicating that children with ADHD have deficient inhibition that is specific to them and that ADHD occurring in the presence of ADHD + CD seemed to represent a phenocopy of CD rather than a variant of ADHD. In other words, individuals with conduct disorder may behave in a way that would lead to a diagnosis of ADHD. Please note that inhibition in this context could also be viewed as an inability to shift or an inability to alter salience calculations. The authors note that the research focus at that time was on the higher-order executive functions due to the fact that the subroutines such as shifting were hard to parse out.

It is also possible that the inability to sustain an appropriate task set in young adults with combined type ADHD, associated with decreased activity in left DLPFC, might in part be due to poor working memory (WM) ability (Burgess et al., 2010). This was considered likely as both processes rely on similar brain areas. Specifically, research has identified the bilateral dorsolateral prefrontal cortex (DLPFC), bilateral inferior frontal gyrus, dorsal anterior cingulate cortex (dACC), inferior parietal lobule, and precuneus as part of the network supporting working memory. These same regions show atypical functioning in individuals with ADHD (Dickstein et al., 2006). They note that lower WM ability in non-ADHD individuals is associated with poorer cognitive ability and increased mind-wandering away from challenging activities in real-life settings.

Recently, specific elements of attentional control have been studied and have suggested specific networks of control for attention control subroutines. For example, right posterior parietal cortex (r-PPC) in ADHD attentional functioning has been studied, with results indicating that activation of r-PPC improves attention-orienting but not alerting and executive control networks in ADHD. In addition, activation of r-PPC worsens selective attention and top-down attentional control in

ADHD, and activation of r-PPC does not affect shifting attention ability and response inhibition in ADHD (Salehinejad et al., 2020). In sum, their findings indicated a domain-specific involvement of the r-PPC in the attention-orienting network of ADHD children. Presumably, future research will find other domain-specific network functions for other subroutines involved in attention and ADHD.

A specific network responsible for shifting attention has been identified (Peelan et al., 2004). This network consists of the premotor cortex, posterior parietal cortex, medial frontal cortex, and right inferior frontal cortex. It responds to both bottom-up and top-down based stimuli.

To show both the complexity and specificity of these networks, adding an auditory cue to signal the necessity to shift from one visual cue to another changed the composition of the network (Salmi et al., 2009). Specifically, they found that an auditory cue to shift visual attention activated the right temporo-parietal junction (TPJ), posterior parts of the left inferior/middle frontal gyrus (IFG/MFG), ventromedial parts of the superior parietal lobule (SPL), and left frontal eye field/premotor cortex (FEF/PMC). In addition, when the cue signaled for a stimulus to be ignored, there was enhanced activity in the ventromedial prefrontal cortex (VMPFC), which the authors considered related to the evaluation of the distracting event. Top-down controlled cue-guided attention shifts (CASs) activated bilateral areas in the SPL, intra-parietal sulcus (IPS), FEF/PMC, TPJ, IFG/MFG, and cingulate/medial frontal gyrus, and crus I/II of the cerebellum. Their results suggested that in audition, top-down controlled and bottom-up triggered shifting of attention activate largely overlapping temporo-parietal, superior parietal, and frontal areas. As the IPS, superior parts of the SPL, and crus I/II were activated specifically by top-down controlled attention shifts, and the VMPFC was specifically activated by bottom-up triggered attention shifts.

The point of all this is that, when it comes to attention control processes, there are varying types of disfunction of attention in general and varying types of attentional disruption in ADHD specifically. These disruptions occur in highly specific neural networks that subserve the processes and subroutines of attention control. They are not the same genetically or in the manner of their white matter connectivity. This implies that these issues should be studied separately. We are also again faced with the question: do these manifestations of attention represent a unified construct? Should they perhaps be designated as discrete entities as neuroimaging advances allow for greater understanding and neuropsychology can better correlate respondent profiles with network processing issues?

Clinical Disorders and Attention Control

There are examples of disorder-specific deficits associated with specific clinical conditions. For example, Gooding, Braun, and Studer (2006) assessed the efficiency of the segregated alerting, orienting, and executive control networks by measuring how response latencies were influenced by alerting cues, spatial cues, and flanking stimuli in individuals with schizophrenia. Visual Flanker tasks are typically composed of a target stimulus in the center, which is flanked by one or more distractor stimuli on either side. Depending on the version of the task being utilized, one selection response might be more correct than another. The finding was the main effects of cue condition and flanker congruency. Their data also revealed a significant interaction of group and flanker type, with schizophrenia-spectrum patients taking longer on average to resolve conflict. They concluded that their results suggested that, compared to the healthy nonpatients, schizophrenia-spectrum patients showed a deficit in their executive control network. They considered that their findings supported the hypothesis that schizophrenia-spectrum patients have a specific attentional deficit rather than a global one. Specifically, their results suggest that patients with schizophrenia have altered alerting abilities. Subsequent research supported this finding (Amado et al. 2011). Interestingly, as far as treatment was concerned for the specific deficit, providing orienting and alerting cues interactively improved attention performance in the resolution of conflict, creating possibilities for cognitive remediation strategies.

Similarly, in a visual search task designed to activate/deactivate the cortical components of the attentional control system in individuals with schizophrenia, Arkin et al. (2020) evaluated the operation of the dorsal and ventral attention networks, lateral prefrontal regions in the frontoparietal network, and cingulo-opercular/salience networks), along with resting state functional connectivity. They found that individuals with schizophrenia exhibited decreased functional connectivity within late visual components and between prefrontal and other components. They also indicated that performance correlated with the deactivation of the ventral attention network in schizophrenia only. This relationship is mediated by the functional connectivity of critical components of the attentional control system. They summarized their results by suggesting that the attentional control system is potentially used to compensate for visual cortex deficits. Furthermore, prefrontal deficits in schizophrenia likely interfered with this compensatory use of the attentional control system. In sum, there are unique attention control deficits in schizophrenia that may require specific and different remediation strategies than those used for attention deficit hyperactivity disorder.

Other clinical disorders display unique problems with aspects of the attention control system. For example, there is evidence to suggest that difficulties in shifting sets may be a precursor to the development of obsessive-compulsive disorder (Bora, 2020). Other research has demonstrated that individuals with obsessive-compulsive disorder have anomalous resting-state large-scale intrinsic networks of cingulo-opercular salience network (SN), default mode network (DMN), and front-parietal network (FPN) (Tomiyama et al., 2022). Interestingly, OCD patients showed reduced functional connectivity between the posterior parietal cortex and dorsolateral prefrontal cortex, which was negatively correlated with anxiety and depressive symptoms within the OCD group. The authors posit that these alterations may represent endophenotype markers of OCD.

Oftentimes, patients with major depressive disorder complain of concentration difficulties that negatively impact their day-to-day function, and these attention problems are not alleviated by current first-line treatments. In spite of attention's influence on many aspects of cognitive and emotional functioning and the inclusion of concentration difficulties in the diagnostic criteria for major depressive disorder, the emphasis of depression as a disease is typically on mood features, with attentional features considered less of an imperative for research or direct treatment (Keller et al., 2019). Nevertheless, deficits of attentional functioning have been noted in depression as well. Specific neural characteristics of individuals with family history of major depressive disorder been identified (Lisiecka et al., 2013). Individuals with major depressive disorder were found to have reduced volume of the right hippocampus, the dorsolateral prefrontal cortex, and the putamen. In addition, adolescents with a family history of major depressive disorder evidenced altered activation in the amygdala and nucleus accumbens when observing emotion-laden pictures of faces (Monk et al., 2008). Importantly, these neural differences were discovered in spite of the absence of negative bias on a behavioral level, suggesting that susceptibility and associated changes can manifest themselves without behavioral signals.

As we have seen, neuroimaging studies using various tasks of attention have identified a common network, often referred to as the frontoparietal attention/control network. This network, which includes areas such as the frontal eye fields, intraparietal sulcus, medial prefrontal cortex (mPFC), and superior parietal lobule, appears to be critical for the control of goal-directed attention, with increasing involvement for more difficult tasks. Research has directly linked hypoconnectivity within this frontoparietal attention network to poorer goal-directed attention in major depressive disorder (Keller et al., 2019).

Clinical depression has been found to be characterized by medium-sized increases of attention maintenance for dysphoric and medium-sized decreases for positive stimuli compared to individuals without depression. Both of these alterations represented what was considered equally important targets for attention modification programs. In addition, individuals who are depressed appear not to manifest abnormalities in early orienting to emotional stimuli. This suggests that differences between patients and healthy subjects in attention to positive stimuli may diminish with age (Suslow et al., 2020).

There is research that indicates that there are specific attentional subroutines of governing attentional processes in different subdomains of various clinical disorders. As an example, among patients with major depressive disorder, Wang and Theeuwes (2020) looked at the specific deficits of attention associated with different subdomains of depression. In a meta-analysis, they found that significant deficits of global attention were observed. Specifically, individuals with major depressive disorder showed significant deficits in psychomotor speed/attention, auditory attention, visuospatial attention, and sustained visual attention, ranging from small to moderate. No deficits in selective attention were indicated.

In summary, attention deficiencies are not specific to attention deficit hyperactivity disorder. They are an inherent component of many clinical disorders. These attentional deficiencies may not respond in the same way to the same treatments that impact attention deficit hyperactivity disorder, and they deserve to be investigated in their own right to determine the most efficacious course of treatment.

Attentional Bias Training (ABM)

This chapter is primarily concerned with attention. However, throughout this chapter, we have spoken about several clinical disorders and how they respond with attention mechanisms, and how they may also respond based upon external and internal variables. For example, let us reflect for a moment about the data regarding attention and depressive disorders. How often may we be missing depression and assigning a diagnosis of ADHD because of the observable deficits in global attention? Are we attending to the overemphasis on threat inherent across multiple pathologies? Are we attending to attentional biases?

We have spoken about disorder-specific remedial approaches to address the deficits found in several clinical conditions in various chapters of this book. We have suggested that there may be clinical interventions designed to remediate these attentional deficits that fall outside those typically used to address the attentional problems found in ADHD. One such example is attentional bias training. This approach addresses the anxiety vulnerability and dysfunction that characterize the attentional bias related to threat. Cognitive training procedures designed to modify selective attentional responding to threats were developed to assess the effectiveness of altering attentional bias as it relates to altering an individual's disposition to respond to a perceived threat (MacLeod et al., 2002). Research has indicated that when ABM training has successfully reduced attention to threat, it also has reduced anxiety vulnerability and symptomatology with a high degree of reliability (MacLeod & Clarke, 2015).

Specifically, MacLeod et al. (2002) developed a computerized task to directly modify attentional biases (AB) for negative stimuli in anxiety. Attentional bias modification training (ABM) was based on a visual-probe task which measured AB. ABM in anxiety had been found to modify relatively early attention processes, such as the speeded detection of disorder-related stimuli (words or pictures), as well as relatively late attentional processes, such as the difficulty to disengage from these stimuli. Typically, the effectiveness of this ABM procedure has been assessed by measuring effects on AB for new stimuli that have not been used in the training (generalization) and on other disorder-relevant cognitions and behaviors. ABM has also been used to successfully treat the deficits of attention symptoms associated with clinical depression (Yang et al., 2015).

This is not to imply that we need disorder-exclusive treatments for the attentional deficit associated with various clinical conditions. In fact, many of the treatments used to improve attention in ADHD are effective in clinical populations as well (Keller et al., 2019). This is particularly true for psychopharmacological interventions. What we do suggest is that there may be targeted treatments for specific disorders that are currently being overlooked or underutilized. Hopefully, as we gain a greater understanding of how neural networks become engaged and under what circumstances, providing greater specificity in our treatments will become the norm.

References

Amado, I., Lupiañez, J., Chirio, M., Landgraf, S., Willard, D., Jean-Pierre Olié, J. P., & Krebs, M. O. (2011). Alertness can be improved by an interaction between orienting attention and alerting attention in schizophrenia. *Behavioral and Brain Functions, 7*, 1–9.

Arkin, S. C., Ruiz-Betancourt, D., Jamerson, E. C., Smith, R. T., Strauss, N. E., Klim, C. C., ... & Patel, G. H. (2020). Deficits and compensation: Attentional control cortical networks in schizophrenia. NeuroImage: *Clinical, 27*, 102348.

Barker, H., Munro, J., Orlov, N., Morgenroth, E., Moser, J., Eysenck, M., & Allen, P. (2018). Worry is associated with inefficient functional activity and connectivity in prefrontal and cingulate cortices during emotional interference. *Brain and Behavior, 8*(12), e01137. https://doi.org/10.1002/brb3.1137

Benes, F. (2000). Emerging principles of altered neural circuitry in schizophrenia. *Brain Research Review, 31*(2–3), 251–289.

Berggren, N., & Derakshan, N. (2013). Attentional control deficits in trait anxiety: Why you see them and why you don't. *Biological Psychology, 92*(3), 440–446. https://doi.org/10.1016/j.biopsycho.2012.03.007

Bora, E. (2020). Meta-analysis of neurocognitive deficits in unaffected relatives of obsessive-compulsive disorder (OCD): comparison with healthy controls and patients with OCD. *Psychological Medicine, 50*(8), 1257–1266. https://doi.org/10.1017/S0033291720001634

Braver, T. (2012). The variable nature of cognitive control: A dual mechanisms framework. *Trends in Cognitive Sciences, 16*(2), 106–113. https://doi.org/10.1016/j.tics.2011.12.010

Burgess, G. C., Depue, B. E., Ruzic, L., Willcutt, E. G., Du, Y. P., & Banich, M. T. (2010). Attentional control activation relates to working memory in attention-deficit/hyperactivity disorder. *Biological Psychiatry, 67*(7), 632–640.

Cieslik, E., Mueller, V., Eickhoff, C., Langner, R., & Eickhoff, S. (2015). Three key regions for supervisory attentional control: Evidence from neuroimaging meta-analyses. *Neuroscience & Biobehavioral Reviews, 48, 10.10*, 22–34.

Corbetta, M., & Shulman, G. (2002). Control of goal-directed and stimulus-driven attention in the brain. *Nature Reviews Nueuroscience, 3*, 201–215.

Coull, J., Sahakian, B., & Hodges, J. (1996). The alpha (2) antagonist idazoxan remediates certain attentional and executive dysfunction in patients with dementia of frontal type. *Psychopharmacology (Berlin), 123*(3), 239–249.

Derryberry, D., & Reed, M. (2002). Anxiety-related attentional biases and their regulation by attentional control. *Journal of Abnormal Psychology, 111*(2), 225–236. https://doi.org/10.1037/0021-843X.111.2.225

Dickstein, S., Bannon, K., Castellanos, F., & Milha, M. (2006). The neural correlates of attention deficit hyperactivity disorder: An ALE meta-analysis. *Journal of Child Psychology and Psychiatry, 47*, 1051–1062. https://doi.org/10.1111/j.1469-7610.2006.01671.x

Dosenbach, N., Fair, D., Cohen, A. L., Schlaggar, B., & Petersen, S. (2008). A dual-networks architecture of top-down control. *Trends in Cognitive Sciences, 12*(3), 99–105. https://doi.org/10.1016/j.tics.2008.01.001

Eysenck, M. (1992). *Anxiety: The cognitive perspective*. Erlbaum.

Eysenck, M. W., & Calvo, M. (1992). Anxiety and performance: The processing efficiency theory. *Cognition and Emotion, 6*, 409–434.

Eysenck, M., Moser, J., Derakshan, N., & Hepsomali, P. A. (2022). A neurocognitive account of attentional control theory: How does trait anxiety affect the brain's attentional networks? *Cognition and Emotion, 37*(2), 220–237. https://doi.org/10.1080/02699931.2022.2159936

Gooding, D. C., Braun, J. G., & Studer, J. A. (2006). Attentional network task performance in patients with schizophrenia–spectrum disorders: evidence of a specific deficit. *Schizophrenia Research, 88*(1–3), 169–178.

Keller, A., et al. (2019). Paying attention to attention in depression. *Translational Psychiatry, 9*, 279. https://doi.org/10.1038/s41398-019-0616-1
Lisiecka, D. M., Carballedo, A., Fagan, A. J., Ferguson, Y., Meaney, J., & Frodl, T. (2013). Recruitment of the left hemispheric emotional attention neural network in risk for and protection from depression. *Journal of Psychiatry and Neuroscience, 38*(2), 117–128.
MacLeod, C., & Clarke, P. (2015). The attentional bias modification approach to anxiety intervention. *Clinical Psychological Science, 3*(1), 58–78. https://doi.org/10.1177/2167702614560749
MacLeod, C., Rutherford, E., Campbell, L., & Ebsworthy, G. (2002). Selective attention and emotional vulnerability: Assessing the causal basis of their association through the experimental manipulation of attentional bias. *Journal of Abnormal Psychology, 11*, 107–123.
Makovac, E., Fagioli, S., Rae, C. L., Critchley, H., & Ottaviani, C. (2020). Can't get it off my brain: Meta-analysis of neuroimaging studies on perseverative cognition. *Psychiatry Research: Neuroimaging, 295*, 111020. https://doi.org/10.1016/j.pscychresns.2019.111020
Monk, C., Klein, R. G., Telzer, E. H., Schroth, E. A., Mannuzza, S., Moulton, J., III, & Ernst, M. (2008). Amygdala and nucleus accumbens activation to emotional facial expressions in children and adolescents at risk for major depression. *American Journal of Psychiatry, 165*(1), 90–98.
Pardo, J., & Janer, K. R. (1990). The anterior congulate cortex, mediates processing selection in the Stroop attentional conflict paradigm. *PNAS, 86*(1), 256–259.
Peelan, M., Heslenfeld, D., & Theeuwes, J. (2004). Endogenous and exogenous attention shifts are mediated by the same large-scale neural network. *NeuroImage, 22*(2), 822–830. https://doi.org/10.1016/j.neuroimage.2004.01.044
Posner, M. I., & Boies, S. J. (1971). Components of attention. *Psychological Review, 78*(5), 391.
Posner, M., & Petersen, S. (1990). The attention system of the human brain. *Annual Review of Neuroscience, 13*(1), 25–42. https://doi.org/10.1146/annurev.ne.13.030190.000325
Salehinejad, M., Ghayerin, E., Nejati, V., & Yavari, F. N. (2020). Domain-specific involvement of the right posterior parietal cortex in attention network and attentional control of ADHD: A randomized, cross-over, sham-controlled tDCS study. *Neuroscience, 444*, 149–159. https://doi.org/10.1016/j.neuroscience.2020.07.037
Salmi, S., Rinne, T., Koistien, S., Salonen, O., & Alho, K. (2009). Brain networks of bottom-up triggered and top-down controlled shifting of auditory attention. *Brain Research, 1286*, 155–164. https://doi.org/10.1016/j.brainres.2009.06.083
Salmi, J., Salmela, V., Salmo, E., Mikkola, K., Leppämäki, S., Tani, P., & Hokkanen, H. (2018). Out of focus—Brain attention control deficits in adult ADHD. *Brain Research, 1692*, 12–22. https://doi.org/10.1016/j.brainres.2018.04.019
Schachar, R., Mota, V. L., Logan, G. D., Tannock, R., & Klim, P. (2000). Confirmation of an inhibitory control deficit in attention-deficit/hyperactivity disorder. *Journal of Abnormal Child Psychology, 28*, 227–235.
Seeley, W. (2019). The salience network: A neural system for perceiving and responding to homeostatic demands. *Journal of Neuroscience, 39*(50), 9878–9882. https://doi.org/10.1523/JNEUROSCI.1138-17.2019
Sturm, W., & Willmes, K. (2001). On the functional neuroanatomy of intrinsic and phasic alertness. *NeuroImage, 14*, 76–84.
Suslow, T., Hußlack, A., Kersting, A., & B. C. (2020). Attentional biases to emotional information in clinical depression: A systematic and meta-analytic review of eye tracking findings. *Journal of Affective Disorders, 274*, 632–642. https://doi.org/10.1016/j.jad.2020.05.140
Tomiyama, H., Murayama, K., Nemoto, K., Tomita, M., Hasuzawa, S., Mizobe, T., ... & Nakao, T. (2022). Increased functional connectivity between presupplementary motor area and inferior frontal gyrus associated with the ability of motor response inhibition in obsessive–compulsive disorder. *Human Brain Mapping, 43*(3), 974–984.
Wang, B., & Theeuwes, J. (2020). Salience determines attentional orienting in visual selection. *Journal of Experimental Psychology: Human Perception and Performance, 46*(10), 1051–1057. https://doi.org/10.1037/xhp0000796

Weber-Goericke, F., & Muehlhan, M. (2019). A quantitative meta-analysis of fMRI studies investigating emotional processing in excessive worriers: Application of activation likelihood estimation analysis. *Journal of Affective Disorders, 243*, 348–359. https://doi.org/10.1016/j.jad.2018.09.04

Yang, W., Ding, Z., Dai, T., Peng, F., & Zhang, J. (2015). Attention Bias Modification training in individuals with depressive symptoms: A randomized controlled trial. *Journal of Behavior Therapy and Experimental Psychiatry, 49*(Part A). https://doi.org/10.1016/j.jbtep.2014.08.005

Chapter 8
Attention and the Role of the Reward Recognition Network

Attention as a concept can be traced back to the eighteenth century in the field of philosophy, thereby predating the foundation of psychology. Within the field, one such philosopher, Gottfried Leibniz, defined the concept "apperception" as necessary for information to enter conscious awareness. Perhaps, consistent with the times, he attributed this awareness as follows: "Attention is a determination of the soul to know something in preference to other things" (Andrade, 2023). A major theoretical advance came with the idea that attention could be externally directed as in processing sensory stimuli, or internally directed such as in reflecting on one's own thoughts (James, 1890). In the latter nineteenth century, attention entered the field of study in psychology under Wilhelm Wundt and Hermann von Helmholtz (Andrade, 2023) and was largely measured based upon directly observable behavior in response to orienting to sensory stimuli. Still, more recently, neuroimaging techniques have allowed us to better assess attention mechanisms or, more specifically, the role of various neural pathways as we attempt to measure various constructs of attention, including the brain during exposure to sensory stimuli and during states of self-reflection. In other words, neuroimaging allows us to see how well or poorly these functional behavioral constructs overlay neural networks. In fact, while neuroimaging has confirmed various attention-related networks, the neural network operations do not necessarily reflect the behaviorally defined constructs that comprise the standard clinical modeling of attention and attention deficits. The answer is then that, in the clinical realm, the functional properties of these now observable neural networks do not necessarily reflect or coordinate well with the behavioral definitions of attention, or the behaviorally defined disorder we call attention deficit disorder.

What we now know with certainty is that humans are active processors. Whether this is for survival or enjoyment, our ability to detect and process, or move toward a selected item, area, thing, or concept is based upon multiple factors including our noticing of the object (arousal), based upon the salience of the sensory stimulus,

T. Wasserman, L. D. Wasserman, *Disorders of Attention*, Neural Network Model: Applications and Implications, https://doi.org/10.1007/978-3-031-78732-4_8

how much value we have assigned to it, the effort required and the resulting motivation. And, if we are to understand a truly complex construct such as attention in terms of its operation over neural networks, then it is imperative to clarify how combinations of neural networks function to allow a human to select for pursuit, a specific environmental stimulus from all of the available stimuli for potential action. In fact, as implied above by the mention of arousal, value, effort, and motivation, it is clear that it is imperative to keep in mind that attention requires the interconnectedness of multiple network systems rather than just one dedicated network as has sometimes been portrayed. That is, attention may bring online, anatomically separate but functionally related neural pathways in order to achieve the symphony of factors necessary to create attention and then sustain attention.

Neuroimaging, diffusion tensor imaging (DTI), and functional magnetic resonance imaging (fMRI) have been invaluable in establishing the presence of large-scale networks, that is, white matter tract connections that run in cortical–cortical and cortical–subcortical loops, which has helped to highlight the fact that there are multiple networks involved in varying aspects of attention and related constructs. There is recognition of several large-scale networks, which we will review. For example, one such network is known as the "default mode network" and is described as a network that spans multiple cortical regions, incorporating parts of the prefrontal, parietal, and temporal cortices and that shows joint activation, or deactivation, in association with particular, more introspective mental functions and has been put forth as confirmation of one such mental function as an element of attention we might think of as daydreaming or being on "autopilot," or attention turned inward. This in turn leads to the observation/confirmation that there are in fact differing elements of attention as characterized by different combinations of networks, or differing combinations of networks reflecting differing elements of attention.

In this chapter, however, we will largely discuss the operation of one of these networks, the reward recognition network, and its role in certain types of attentional processes. It is probably safe to say that in the clinical setting, reward does not frequently come up as attention is being discussed. However, we have identified this network as a key component in motivation and human behavior (Wasserman & Wasserman, 2020) and much of what we will mention here has been discussed in alternative writings by the current authors in relation to understanding how human motivation operates within a network model. We will see how there is considerable overlap between these constructs of reward and motivation, and to some extent, how they cannot be completely dissociable from each other when looking at everyday human behavior. After all, it is hard to be motivated by something to which you have not first paid attention.

Finally, it is important to understand that attention is directly tied to motivation and many networks are shared by these two constructs. As we explore the reward system, it will be difficult to escape the conclusion that attention is a necessary precondition to motivation, based in large part, on reward and may not be dissociable from it in real-life practice.

However, in order to address reward and motivation, we must attend to a precursor of motivation, i.e., attention. We will focus on the origins of attention from both a developmental perspective and a specific object/goal-related factor and end with a discussion of how disorders of attention are implicit in both neurodevelopmental and mental health disorders. We will take a look at these large-scale networks and how some elements of attention are associated. We will then focus on a review of what is known about the reward recognition network and why it is important when considering what can go wrong with attentional processes.

As a starting point, it is important to start at the beginning. Let us step away for a moment from the concepts and constructs of attention and motivation as we understand them in their fully developed or adult forms. That is, before the brain represents a mature neurodevelopmental organ both physiologically and as a result of extensive experience. Let us instead think for a moment about their development.

The Development of Attention

When we think of a newborn, we are now looking at an organism that does not yet have a learned history. As such they do not have many examples of what has previously "caught their attention" and that which they have found rewarding. Rather, they are about to engage the environment in what is known as "bottom-up" processing. This is a process driven by external sensory stimuli for which we do not have prior knowledge. Think of it as a foundation upon which we continue to build based upon experience. In preparation for this emergent processing, we have what the current authors have relabeled, from primitive to foundational reflexes, which help us to adapt to our extrauterine environment. These foundational systems are initiated during gestation (Wasserman, 2023) and the foundational reflexes in the extrauterine environment serve to orient us to what is salient for our survival and to guide us later on to the exploration of our environment, serving as the foundation upon which all further motor and cognitive development occurs.

The "orienting response" was first coined by Pavlov (1927) and spoke to the newborn's innate reflex to orient to stimuli in their environment. This process has been described as a "bottom-up" cognitive process wherein we attend to novel stimuli and broaden our knowledge and experience base. As we develop, as we gain experience, this "bottom-up" process is replaced by a "top-down" process, which is guided by, in part, reward history. Thus, while our initial attention is driven by an innate reflex, designed as a survival mechanism, it later becomes moderated by experience as we attend to that which we find rewarding, or of value. It is then the foundational precursor to instrumental attention.

For the neonate, most responses to the environment are based upon adaptive foundational reflexes. We orient to that which will address our survival needs. We learn the value of these extrinsic stimuli and then begin to orient to them with intent as we store the stimuli we have determined to be of value in memory. We learn to locate them and respond quickly to their presence because they have assumed value. Simultaneously, we also learn to suppress the intensity of our attention to extraneous environmental stimuli that are routinely present in order to reserve that cognitive energy for the new or novel stimuli that will require or, for other reasons, garner our attention. That is, we do not necessarily orient or pay attention to the routine, either because we have mastered that routine or because it is a routine in our environment that poses no threat. We inhibit that response, meaning we do not use our limited attention resources to attend to that which is routine or not of consequence to our survival.

Although we will discuss this in much greater depth, we wish to take this opportunity to draw the reader's attention to the beginning of the implications for what is currently termed an attention deficit disorder. We would argue that it is not an attention deficit per se, but rather a failure to allocate attention appropriately. It may be a failure to mask extraneous environmental sounds or internal cognitions, but it is not a deficit of attention per se. For example, if one is aphasic, one will always have difficulty with language, which is verbal expression or reception. Persons who have been labeled with the neurodevelopmental disorder ADHD *can* pay attention, provided the conditions are in alignment. The better question is: what is that child or person attending?

Returning to our immediate topic, the process of development from a biological and psychological developmental perspective is the scaffold for neural network development. From a developmental perspective, these networks begin highly confined in both function and structure. As we progress, so does the level of sophistication of the networks, from separate small world hubs to large-scale networks that reach further and with greater integration of network hubs. We know this as cerebral integration from a neurophysiological perspective which possibly aids in the development of executive function from a cognitive psychology perspective. While these involved networks can be differentiated in many ways, it is important to remember that network associations are unique, and networks come online to solve a particular problem occurring at a particular time.

A word about networks—Before we begin our discussion of the reward recognition network, we provide a brief description of the major neural networks of the human brain.

Large-Scale Networks

The brain is a network of highly interconnected regions. With the introduction of functional magnetic resonance imaging (FMRI) and diffusion tensor imaging (DTI), functional network communities have been identified. These neuroimaging techniques utilize various methods to observe changes in blood, water flow, and activity strength to determine the anatomical substrates of functional constructs. In addition to the availability of various techniques, it is important to note that researchers may use slightly different measures for determining activity, thereby resulting in slightly different areas to be included, or not, within a given network, resulting in some variations in the anatomical and functional labeling of these networks. Resting state/always on" networks refer to the analysis of functional brain connectivity in the state of rest. Resting state fMRI (rs-fMRI or R-fMRI) is a method of functional magnetic resonance imaging used to determine networks or functional communities called resting state networks; the brain mapping of regional interactions that occur in a state when no external task is presented. These fMRI methods have uncovered different resting state networks, which represent specific functions and varied spatial topology. A number of resting state brain networks have been identified, including the aforementioned default mode network. Resting state/"always on" networks in the brain that have been identified include the salience network, auditory network, basal ganglia network, higher visual network, visuospatial network, default mode network, language network, central executive network, precuneus network, primary visual network, and sensory motor network. This is a good time to remember that the coming online of neural pathways is highly task dependent. A summary of these networks is given as follows.

The Default Mode Network

The default mode network is activated when a person is not actively engaged or when a person is not focused on what is happening around them. It hypothetically represents the internal modes of cognition used when remembering, thinking about the future, and mind wandering. It is also known as the task-negative network because it becomes deactivated when the individual engages in some tasks. The default mode network involves the posterior cingulate cortex, medial prefrontal cortex, and lateral parietal cortex.

This network is otherwise known as the mentalizing network due to its participation in social cognition such as introspection, mind wandering, emotional processing, thinking others' mental state, and so on (Fan et al., 2021; Seeley, 2019; Smallwood et al., 2021). As a part of the default mode network, the precuneus is a brain region involved in a variety of complex functions. These include memory and memory integration of information concerning perception of the environment, cue reactivity, mental imagery strategies, episodic memory retrieval, and affective responses to pain (Science Direct, 2020).

Salience Network

Menon (2015) talks about the salience network as anchored in the anterior insula and dorsal anterior cingulate cortex and is a paralimbic–limbic network. This network also includes nodes in subcortical regions of the amygdala, hypothalamus, ventral striatum, thalamus, and specific brainstem nuclei (Seeley, 2019), thus bridging the emotional and cognitive systems. As noted above, definitions of nodal boundaries may vary across researchers. The salience network is an integral part of other brain networks and functions. The salience network has a central role in the detection of behaviorally relevant stimuli and the coordination of neural resources. It is involved in detecting and filtering salient stimuli, and in the recruitment of relevant functional networks. Simply put, this network has a key role in identifying what target to respond to, or what behavior to perform next. This network is also activated when threats are perceived or novelty encountered. It may be involved in the internal feelings associated with reward (Pace-Schott, 2017) or emotional engagement (Menon, 2015). Together with these interconnected brain networks, the salience network is essential to a variety of complex functions, including social behavior, communication, and self-awareness through the integration of sensory, emotional, and cognitive information across networks.

Disruption of the salience network will disrupt the functioning of other networks because it has a key role in regulating the dynamic changes in other networks. Moreover, the network is essential during periods of rapid change of behavior. That is, what to do next or not to do next is decided by the appropriate functioning of the network. For these reasons the salience network's proper functioning is invaluable for the commencement of control of cognition processes. Salience Network dysfunction has been observed in various psychiatric disorders including anxiety and schizophrenia (Menon, 2011), among others. For example, Menon (2015) notes one example of disruption of the salience network as related to autism, wherein an individual may demonstrate poor social skills because of the diminished salience of social stimuli such as facial cues.

The Central Executive Network and Executive Control Network

We have written about the lack of a precise definition of the term executive function (Wasserman & Wasserman, 2013). For the purposes of this discussion, let us define executive function as referring to a set of cognitive skills that underlie the capacity to plan ahead and meet goals, display self-control, follow multiple-step directions even when interrupted, and stay focused despite distractions. The network is constituted by the dorsolateral prefrontal cortex and posterior parietal cortex. The medial frontal gyrus, superior frontal gyrus and the anterior cingulate cortex, paracingulate gyri, ventrolateral prefrontal cortex, and subcortical regions of the thalamus constitute the executive control network. This network is active during a task that needs cognitive control and working memory. The executive control network is considered to be functioning during target-directed activities and during control of intellectual activities (Dong et al., 2015).

There are neuroimaging studies to demonstrate that networks may be defined, but may also work in unison. Specifically, in an effort to substantiate earlier work, Goulden et al. (2014) focused on the dynamics of switching between the default mode network (DMN), the network that is active when the brain is not specifically engaging in a task, and the central executive network (CEN), which is active when the brain is engaging in a task that does require attention. They successfully demonstrated that the salience network, which responds to the subjective salience of a stimulus, drives the switching between the default mode network and the central executive network. This serves to highlight the interconnectedness functionally of separate networks. It also helps to tie together the point that salience moves us from a more passive processing to active processing.

Auditory Network

This network comprises the brain regions necessary for the processing of auditory information. The nodes associated with the auditory network are the right and left primary auditory cortex, Heschl's gyrus, planum polare and temporale, lateral superior temporal gyrus, and posterior insular cortex. The auditory cortices have been functionally well defined as reliable to the anterolateral Heschl's gyrus, left and right, respectively (Andoh et al., 2015).

Basal Ganglia Network

The basal ganglia refer to a group of subcortical nuclei responsible primarily for the regulation of behavior. This network assists with initiation as well as persistence and inhibition. Although originally considered primarily in relation to motor behavior, we have written extensively about the role of this network in most cognitive activities. The network is involved in many functions such as control of motor areas, emotion, cognition, and so on. That is, the basal ganglia are involved in roles such as motor learning, executive functions and behaviors, and emotions. It plays a key role in the learning of difficult and complex behaviors, and due to its strong role of coordination during movement, it is involved in goal-directed behaviors that require movement. It has also been demonstrated that the basal ganglia are sensitive to both positive and negative reward characteristics (Joel, 2000). Specifically, the network involves the basal ganglia, substantia nigra, subthalamic nucleus, striatum, and global pallidum externa and interna. The striatum receives cortical input through the direct and indirect pathways. The input is always excitatory; therefore, the basal ganglia are active only when required, depending upon the nature of the activity or task. We wish to emphasize that the basal ganglia are regulatory. In short, the basal ganglia are involved with both processing sensory data, regulating behavior, including cognition, and the internal state of the organism, such as reward value.

Visual Network

This network is involved in processing information received visually. The visual network consists of nodes in the bilateral and medial calcarine sulcus, extrastriate regions, namely the lingual gyrus, inferior area of precuneus, and lateral geniculate nucleus of the thalamus. The lateral geniculate nucleus of the thalamus functionally connects the visual input to the primary visual cortex. There is a primary visual network consisting of the peristriate area, lateral occipital gyrus, and superior occipital gyrus (Damoiseaux et al., 2006). The visual areas situated in the extrastriate regions of the lingual gyrus, inferior division of precuneus, and lateral geniculate nucleus of the thalamus constitute the medial visual areas. The occipitotemporal junction, spreading out from the occipital pole and the superior parietal region constitutes the lateral visual areas (Beckmann, 2005).

The Visuospatial Network

Visual–spatial attention is a form of visual attention that involves directing attention to a location in space. Spatial attention selectively processes visual information by prioritizing an area within the visual field. Once a region is selected, the information within it receives additional processing. The visual–spatial network consists of the posterior parietal cortex of the occipitoparietal junction, midline of the precuneus, posterior cingulate cortex, and the frontal pole. As can be appreciated by its function, it plays an active role in many attentional processes.

The Language Network

The language network is responsible for language processing. The language network is left lateralized with the major nodes of the language system identified as incorporating both Broca's and Wernicke's areas connected by a massive network of neurons called the arcuate fasciculus. It includes the prefrontal, temporal parietal, and subcortical nodes. The language network is then involved with functions much more complex than simply expressive or receptive speech. Although it is involved in both expressive and receptive language, the connection between the major cortical areas now creates functional association, resulting in involvement of comprehension, reading, interpreting, mimicking, and so on. In addition, Broca's area is the location of mirror neurons where the neuron's firing rate increases when the individual performs goal-directed activities and while observing similar movements of others (Bronco, 2020). Disruption to this network can then result in disruptions of expressive and/or receptive language.

The Sensorimotor Network

The sensorimotor network helps interpret our response to external stimuli as well as somatosensory responses to internal proprioceptive sensations including visceral responses and balance and coordination. The sensorimotor network includes functional areas in the primary motor cortex, cingulate cortex, premotor cortex, and the supplementary motor area, as well as primary and sensory cortices in the parietal lobe. The sensorimotor network is also known as somatomotor network, a large-scale brain network that primarily includes somatosensory (postcentral gyrus) and motor (precentral gyrus) regions and extends to the supplementary motor areas (SMA). The auditory cortex is sometimes included (Sohn et al., 2021). This network allows for the interpretation of sensory information in coordination with motor movement.

The Reward Recognition Network

Identifying a reward is conceptually a critical function in the process of allocating attention. In fact, the study of reward is often broken down into phase: anticipation of reward, responsivity to reward once received, and reward-based learning. The neural circuits most closely associated with reward processing are the ventral tegmental area, ventral striatum, including the nucleus accumbens, which is considered to be central to considering reward value, the anterior cingulate cortex, medial prefrontal cortex, and orbital frontal cortex. There are then a number of potential networks that make contributions to the process of reward recognition, including networks responsible for error predictions and adjustments, emotional responding, memory, and the one responsible for reward identification itself.

Goals, Salience, and History

Historically, the question of how an individual searches the environment, and more generally, how we parse information from the environment, is studied in the context of attentional control. Many models of selective attention have described attentional control as a result of the interplay between voluntary, top-down, or goal-driven control and automatic, bottom-up, or stimulus-driven control. This chapter is about the reward recognition network and its role in attention. Studies demonstrate that attentional selection does not need to be the result of the traditionally identified top-down or bottom-up processing systems we discussed. Instead, it is often driven by lingering preferences due to a memorized "history" of former attention deployments (Failing & Theeuwes, 2018). This memorized history includes the recognition and identification of reward history and the related construct of value. However, selection biases (attention) can be neither explained by current selection goals nor by the physical salience of potential targets. Awh et al. (2012) suggested that a third factor they labeled as "selection history" explains the competition for selection. There is some history attributing this phenomenon to top-down processing. However, Awh et al. posit lingering selection biases formed by the history of attentional deployments that can be unrelated to current top-down goals or the physical salience of items. They propose that directed attention can be determined not only by current goal (top-down) or stimulus-driven (bottom-up) selection but also stronger because of a cumulative effect of priming by the lingering effects of past selection episodes, sometimes in direct conflict with current goal-driven orienting stating.

Neuroeconomics; Value and Attention

There are specific aspects of learning, including which rewards we determine to be of value to us, that are directly related to the development of attention. Since all events have the potential of being rewarding, some more so than others, how do we go through this complex process of determining the "worth" or reward value of any event or object or experience? Furthermore, how does this determination of worth figure into the deployment of attention.

Increasingly, behavioral researchers have conjectured that individuals make choices by first establishing and then assigning values to the stimuli in the environment. These values represent the relative desirability of that particular stimulus as compared to all the other stimuli in the environment (Armel et al., 2006). This concept is useful when considering how a person deploys attention. Several models have been developed that contain the central idea that attention is deployed based on a forecast or prediction of the value of a choice based on a prior interaction with that choice or something similar to it.

One way to understand how attention is deployed is to consider what happens when the determination of worth that is recorded in memory is used by the individual to anticipate worth the next time the same, or similar, stimulus is encountered. This concept of "reward anticipation" includes cue evaluation, motor preparation, and feedback anticipation (Zang et al., 2017). The process itself is also tied to effort expenditure. Yi et al. (2020) found support for the idea that the amount of effort expenditure reduces reward sensitivity during the anticipatory phase and enhances reward sensitivity during the consummatory phase. That is, for any given task or goal-driven behavior, the amount of effort required to secure a goal impacts our attention to the goal, as well as our potential motivation to secure the said goal. Subsequently, attaining the goal successfully increases its valence for identification and selection the next time it is encountered. Long-term memory mechanisms related to the recognition of reward must balance the need to represent potential determinants of all available reward outcomes with the computational burden of an overinclusive memory and limited working memory capacity (Murty & Adcock, 2014).

Remembering that reward value, in a positive mode, would likely increase future efforts in that direction, how then might the process be facilitated. One solution would be to enhance memory for specific salient events that occur during reward anticipation. The memory enhancement would likely also have emotional saliency. This has, in fact, been demonstrated by research, indicating that reward motivation can enhance hippocampus-dependent memory and anticipatory ventral tegmental area (VTA)–cortical–hippocampal interactions (Ranganath, 2010). Other research demonstrates the dopaminergic neuromodulation of prefrontal function and hippocampus-dependent memory (Berridge & Robinson, 1998). Both of these findings have implications for attention as the increased reward saliency increases the value of the target for future encounters.

One possible way to appreciate the role of attention is to recognize how one particular signal becomes more amplified, selected, and behaviorally regulated than others in the immediate stimulus field. This process has been identified as establishing the reward salience of the target. This reward, or incentive salience, refers to motivation for rewards that is driven by both physiological states and previously learned associations about a reward cue (Hyman, 2005). Although discussed most often in the context of motivation, the idea of salience has implications for attention as well. Based on the idea of value, the stimulus with the most historically attached value will have the greatest ability to draw and regulate selection for action (cause the individual to "pay attention"). In sum, once reward value is established, we engage in a constant process of evaluating rewards, reassessing reward valuation, and determining whether or not to engage. Reward salience is therefore an important addition to our understanding of the initiation of arousal and its regulation of attention. It can also be reflective of potential problems when this aspect of the process goes awry, for example in depression.

The concept of saliency has gained increasing recognition in the understanding of clinical disorders such as psychosis (Bossong & Kahn, 2016) wherein findings are of aberrant processing of reward salience and disordered attention. A fuller discussion of this concept and implications for disorders of mental health is included in the chapter on "Saliency."

In addition to reward anticipation and salience, a related candidate is that of error analysis or error prediction, an ongoing process constantly modulating reward value. One way the neural network might operationalize this process is for ventral tegmental area modulation to induce distributed neural changes that amplify hippocampal signals, thereby associating memory with reward value. Another way is for memory to record the impact of expectancy violations, which might modify saliency in a downward trend rendering the stimulus to be determined to be of less reward value. That is, the process involves history, working memory, and attached emotional value with analyses either enhancing or diminishing the reward value of a salient item.

Some Background on Neural Networks and Implications for Attention

Human neural networks represent a major paradigm shift in understanding the structures of the human brain responsible for information processing. The shift represents a movement away from the module-specific understanding of brain regions to one where an understanding of how the interconnectedness of the human brain is involved in most human cognition. In this regard it is very important to remember that neural network modeling includes emotional and personal memory aspects (Wasserman & Wasserman, 2017). Most human processing is not free from "emotional influences" although emotional influences vary from task to task and person to person. For example, it should not be surprising that children who "hate" school are also likely to do less well academically as compared to children who "like" or recognize the benefit of schooling.

Neural networks are the future and past repositories of all that makes us both human and unique. Therefore, while certain classes of tasks may recruit certain cortical structures and regions across persons, as components of their particular network, each network is task specific and, in some way, unique to each person. This can be perhaps better imagined by looking at neural imaging studies across subjects; some common areas are recruited, but scatter or personal variations exist. The neural network combinations involved in any particular task or experience are, however, mappable, with some commonalities and some unique variations.

To reiterate an important concept, at its core, a neural network, or circuit, is simply a population of neurons located throughout the brain that are interconnected through synapses and organized according to small world hub principles. These neurons are interconnected through the brain in a task-dependent manner to carry out certain specific functions. In many cases, these "networks" are not permanent. Each new task that a person faces causes the creation of a neural pathway to come online to solve it. Once solved, the network components often become available to be "recruited" to address other tasks. How we pick the task to solve involves both attention and motivation. Deciding which task to solve involves selecting a particular task from a set of tasks to attend to. Deciding or selecting is based on a number of factors that essentially comprise what we will come to call attention and/or motivation. These include factors based in history such as how much do I like what will happen if I solve this task or engage in this activity, how likely is it that I will actually be able to do what is required, and how much effort will I have to expend to do what is required. It is important to remember that attention does not always imply deciding or top-down regulating. Aspects of attention are automatic based on reinforcement history or experience.

There are, however, tasks and situations that recur frequently. These cause the creation of collections of neurons and structures that have purposes that are constantly being utilized by a functioning brain. These can be recognizable as known networks because the demands of the tasks require skills to solve them that are common or shared across persons. One of these networks is the reward recognition network, and it is utilized to learn about, store, remember, and recall environmental stimuli that were rewarding for the person. That is the essence of what it does. The relationship of this function to a construct like attention has to be established by connecting the function of the network to the operation of the construct. As we have seen above, we can model just such a connection, and begin to discuss how the recognition of rewards contributes to the development of attentional regulation. Examples of positive reward may include our finding of a particular chocolate that we deem superior to all others. While sailing down the supermarket aisle, I might spy just that chocolate as its reward value of that chocolate increases its salience and captures my attention.

Attention and Motivation: What's the Difference?

If you think at this point that it would be easy to conflate attention and motivation, you would not be incorrect. The traditional understanding of attention segmented it into two major components: top-down processing versus bottom-up processing. Bottom-up processing involves automatic, stimulus driven behaviors based on the idea of novelty or instincts as the primary drivers of attention. Top-down processing involves the imposition of cognitive control over stimulus selection for attention. More recently, at least for visual attention, it was argued that in addition to top-down and bottom-up processes, lingering biases of selection history play a major role. Reward salience suggests that experience with selecting a particular feature, or the location of a feature, may boost and sharpen its representation within the priority selection map above and beyond its physical salience. In other words, experience may render a feature or location subjectively more salient. If this is true, purely cognitive, top-down control of visual selection occurs far less often than what is typically assumed. Most of the time, selection is based on experience and history. In many, if not most, instances, over time it becomes fast, automatic, and occurs without much, if any, effort (Theeuwes, 2018). That is responses, the reward determination and salience, through repeated exposure to a familiar stimulus reaches a level of automaticity no longer requiring the determinations of reward value and determination of effort or cost associated with these familiar stimuli. These learned selection biases are based on value assessments and are considered motivating in the sense that they pull the individual to attend to the stimulus with the most value. All of this suggests that, at some point early in the developmental process, reward valuation, motivation, and attention become, for the most part, inexorably intertwined and not clearly separated. This also suggests that there would be disorders of attention associated with poorly developed reward valuation of specific stimuli-impacting motivation. For the current authors, understanding and explaining ADHD within this framework is most clarifying for both clinicians and, say, parents who often assume a characterological willfulness to their child's lack of attention and compliance. For example, rather than the typical reports of lazy, inattentive, and willful behavior, the understanding of disordered attention from a neural network perspective, including a lack of arousal, a lack of salience for behaviors others value, and limited reward value affecting effort and motivation, provides a much more acceptable understanding of a child's difficulties as biophysiologically based rather than willful misconduct. This is addressed in greater detail in the chapter "Attention Deficit Hyperactivity Disorder and Neural Networks."

Finally, the idea of associating value with a stimulus as an essential feature in the assignment of attention strikes at the very idea of attention as a singular construct.

Except for the newborn for whom inborn curiosity is a factor, in order for value to play a role in attention, the object must be associated with past experiences. Clearly, this must happen very quickly in the process as items or actions are selected for further attention. This implies two discrete but interrelated processes.

The first one is purely that of object identification: "That is ice cream" or "That is a dog." This is followed by the decision to employ more attentional resources to obtain the object, or less attentional resources to drop the object from consideration. Take this example: You are out walking your dog and the dog breaks away and runs around the corner. You follow the dog around the corner only to find a park where there are lots of dogs running free. You begin by visually scanning the environment for your dog. This is accomplished by looking for movement, then color, shape, or size among other things. The initial category, i.e., movement, is associated with bottom-up processing while some of the other search criteria are associated with top-down imposition of criteria. The dogs that do not meet search criteria are quickly dropped from consideration and are not attended to further. Suddenly you see a dog with the right shape and your attention is increased. As other associated search criteria are met, you "realize" that it is your dog, and since you wish to keep your pet, you are then motivated to retrieve it. This entire search process takes less than a second per candidate until criteria are met and you "focus" on your dog. Attentional difficulties can happen at any stage of the process, and it is worth considering whether the types of problems an individual might have if the bottom-up phase was disrupted are the same problems you would have if the top-down phase of the reward association/value phase went awry. We will consider these very issues later in this chapter. Finally, it is very worthwhile to consider what started this process in the first place: you lost your dog. You were motivated to direct your attention to the search before you even started it. We posit that for humans, this pre-assignment of direction is the rule rather than the exception. Other than startle experiences, most human attention has valuation components throughout the process.

Network Structures of Reward

Attention then involves analyzing, predicting, or anticipating reward and these operations involve a number of brain networks specifically recruited to meet the processing requirements of the task at hand. Central to them all is the fronto-striatal neural circuit, which is a core component of every operation of the reward system (Haber & Knutson, 2010). This fronto-striatal circuit involves dopaminergic projections from midbrain nuclei (e.g., the ventral tegmental area) to subcortical regions that process the rewarding properties of stimuli (e.g., the ventral striatum, including the nucleus accumbens) to cortical target regions (e.g., the orbitofrontal cortex, medial prefrontal cortex, anterior cingulate cortex). This circuit, when functioning as part of an anticipatory process, is responsible, in part, for a process that includes reward-responsivity, incentive-based learning, assessing probability of reward receipt, prediction error, and goal-directed behavior (Nusslock & Alloy, 2017). In fact, in both the processes of down-regulation or up-regulation, the cellular responses (increases or decreases in responsivity) to a stimulus such as insulin, or drug addiction, have been shown to occur in the brain regions such as the nucleus accumbens (Nestler, 2014), which is a component of the fronto-striatal network in these sorts of situations.

The frontal striatal circuit has been identified as a key component of the networks subserving attention. Perhaps then not surprisingly, deficits in frontal striatal circuitry have been identified in attention deficit disorder (Durston et al., 2011). These same circuits have been identified in the ability of organisms to time and coordinate temporal sequences of events and to select particular aspects of their internal and external environments to which they will attend (Meek & Benson, 2002). Finally, to emphasize a point stated above, frontal striatal circuitry has been identified as part of the neural networks subserving attention, motivation, emotion, and related psychiatric disturbances (Arsten & Rubia, 2012).

Neural Networks, Behavioral Regulation, and Attention

Network regulatory processes, which are often described as executive functions, appear to be both sensitive to external sensory stimuli and can result in epigenetic alterations as a result of the external stimuli. In turn, this epigenetic influence (Lux, 2018) can influence behavior. Lux (2018) proposed a "dual-activation hypothesis," which states that "concerted activity in both central regulatory networks of the HPA axis and developing sensory networks leads to the establishment and consolidation of epigenetic modifications which underlie the long-lasting programming effects of environmental stressors" (p. 639). The result is that epigenetic alterations in neural networks likely have implications for mental and physical health wherein the resulting epigenetic mechanisms control behavioral responses. This is a phenomenon now being well studied as it relates to addiction, depression, and anxiety. Highly intertwined in this neurophysiological community is the concept of reward, and by extension, motivation and attention.

Returning to our discussion of reward processing networks, a number of brain structures along these network lines are thought to be involved in specific aspects of reward processing, specifically attention and related anticipation versus attainment or consumption. Attention and anticipation are highly integrated processes in many instances. Although attention can be purely reactive, such as when an individual reacts to an unexpected event, in most cases, attention is under deliberate top-down control that concerns itself with anticipating events in the environment. This is because directing attention and preparing for action takes time. It is then beneficial to be able to predict where an important event will occur in the environment and direct attention to it in anticipation of it happening (Balkenius et al., 2008). The action of directing attention and predicting when the adoption of an anticipation strategy will be positive is made available in working memory when the related environmental stimuli is encountered. While attention, anticipation, attainment, memory, and consumption all occur as part of a highly integrated process of obtaining and utilizing a reward, the processing requirements, and by extension the neural networks subserving them for each of these "phases" are somewhat different. Arguably, a person with an inability to appropriately anticipate outcomes will look inattentive. The core reward networks would be used for all of these processes, but other networks would be recruited depending on the task demands or the phase of the overall reward process. For example, value-based attentional priority only develops for stimulus features that carry uniquely predictive information about reward (Sali et al., 2014). This finding suggests that mechanisms of reward prediction play an important role in shaping attentional priorities. In fact, a genetic trait related to a form of a gene (DRD2 A1 allele) that prevents the expression of the normal laying down of dopamine receptors in brain reward sites has been associated with deficient attention and also predisposes individuals to be at high risk for addictive, impulsive, and compulsive behavioral propensities. These genetic variants of dopaminergic genes, their reliant networks, and other "reward genes" are important common determinants of reward deficiency syndrome (RDS), and are hypothesized to include ADHD as a behavioral subtype (Blum et al., 2008). In addition, anticipation of rewards, which could be involved in the focusing of attention, activates a broad network, including the medial frontal cortex and ventral striatum. Attainment of reward also activates memory and emotion-related regions such as the hippocampus and parahippocampal gyrus, but not the ventral striatum. Reward consumption, in contrast, is associated with amygdala activation and under some conditions, the thalamus. Therefore, the available research suggests dissociable neural networks for the attention to, anticipation of, and attainment of reward. In addition, there are some findings that the neural mechanisms underlying reward consumption are in fact more modality-specific than those for reward anticipation, and that they are also mediated by subjective reward value (Rademacher et al., 2010). For example, Rademacher found differential activations for social reward versus monetary reward in the attainment phase, with the amygdala and thalamus reacting, respectively. That is, while social stimuli were associated with amygdala activation, monetary rewards were associated with the thalamus. Putting this into perspective, these findings of activation of a broader network during the attention and anticipation phases make sense as the processes at these stages involve a more subjective analysis, involving a wider range of outcomes, with each containing a potential reward value, whereas the category of reward invokes different valuations regulated by different subcortical areas.

Positive Prediction Error: Reward Sensitivity

In a related phenomenon, Kal et al. examined the neural underpinnings of reward anticipation, attainment, and positive prediction error (PPE; unexpected reward gain). Positive prediction error is defined as the attainment of an unexpected gain or reward. As noted above, they found that anticipation activated a broad network, including the medial frontal cortex and ventral striatum. Reward attainment activated memory and emotion-related regions such as the hippocampus and para hippocampal gyrus, but not the ventral striatum. PPE activated a right-dominant fronto-temporo-parietal network. This indicated, as we would suspect, that there is an analysis component to prediction errors designed to correct and modify future analyses.

Finally, Pochon et al., (2002) attempted to test the hypothesis that motivational and cognitive processes are not only linked but also by a specific neural system designed to reach maximal efficiency. Their working memory task during an fMRI showed distinctive brain activation areas in relation with changes in both the cognitive loading and the reward-associated loadings with task performance. They noted that the working memory tasks activated a network including the dorsolateral prefrontal cortex. However, under more conditions, the lateral frontopolar areas were evoked. They conclude that lateral prefrontal areas are organized in a caudo-rostral continuum in relation with increases in executive engagement. They also concluded that reward induces an increased activation in the areas already activated by working memory processing and the medial frontal pole as well, regardless of the level of cognitive engagement required. They postulated that the medial frontal pole played a specific role in monitoring the reward value of ongoing cognitive processes. Lastly, there were areas detected where the signal decreased, including ventral and subgenual prefrontal cortices, when both cognitive demand and the reward increased. They concluded that this deactivation may represent an emotional gating aimed at inhibiting adverse emotional signals to maximize the level of performance. Taken together, these results suggest a balance between increasing activity in cortical cognitive areas and decreasing activity in the limbic and paralimbic structures during ongoing higher cognitive processing. They state that: "Converging data demonstrate two parallel neural networks within the Prefrontal Cortex (PFC); one, including the dorsolateral PFC (DLPFC), involved in working memory (WM) and planning, and the other, including the ventral PFC (VPFC), and to some extent the ventral medial areas (MPFC), associated with reward sensitivity and motivation" (Pochon et al., 2002). Their findings suggest that reward sensitivity may recruit attentional control processes during free choice potentially supporting goal-directed behavior and action selection. They suggest that: "Indeed, to be correctly performed, attentionally demanding cognitive tasks required the subjects to minimize the processing of competing signals such as emotional information or interfering thoughts." In this context, reward sensitivity refers to the degree to which an individual's behavior is motivated by reward-relevant stimuli and is believed to be regulated by a behavioral activation system. It requires active balancing between cognitive output and emotionally laden processing. One might see how a disturbance in the balance could affect many areas of daily functioning.

How then do we tie this body of information regarding the human's ability to recognize and analyze reward to the construct of attention? It is to that issue we now turn our attention.

Network Anatomy of Value-Based Decision-Making in the Human Brain

Neuroimaging studies have identified the neural networks activated by all the components of value estimations and choice behavior in humans. For the purposes of discussion, value can be defined as the estimation of how rewarding a particular choice is perceived.

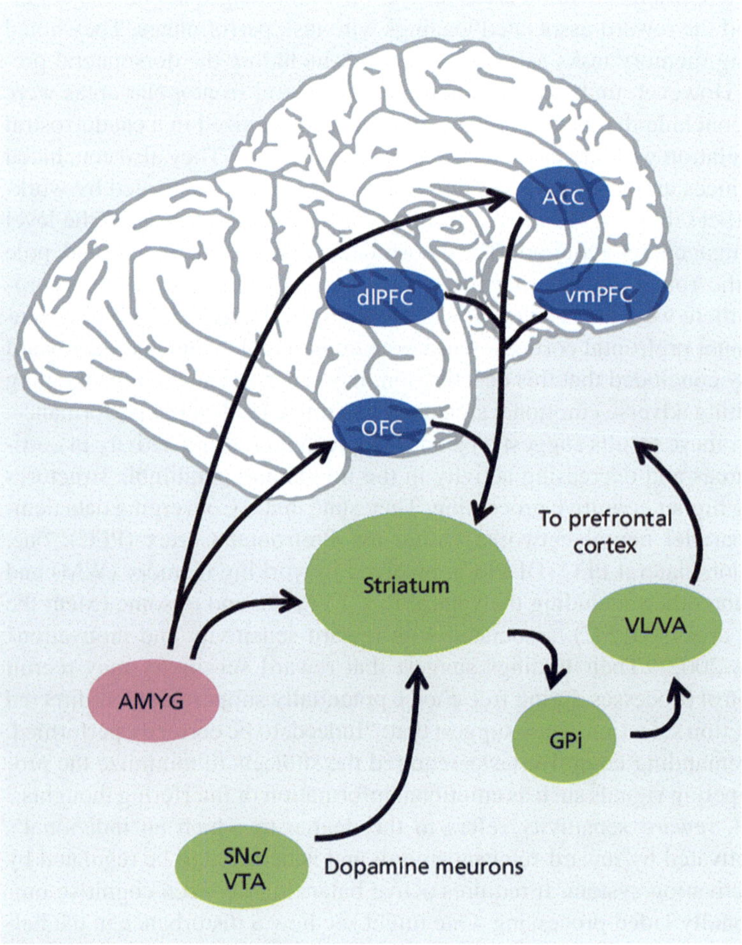

Simplified schematic representation of the reward and decision-making network. Midbrain nuclei containing dopaminergic neurons: AMYG amygdala, GPi internal globus pallidus, SNc/VTA substantia nigra pars compacta/ventral tegmental area, VL/VA ventral lateral and anterior thalamic nuclei. Subdivisions of the prefrontal cortex: ACC anterior cingulate cortex, dlPFC dorsolateral prefrontal cortex, OFC orbitofrontal cortex, vmPFC ventromedial prefrontal cortex (Sirigu & Duhamel, 2016)

Reward-related behavior emerges from the activity of interconnected neural networks rather than from any single brain structure or one specific network (Kelley & Berridge, 2002). As we have seen (Sali et al., 2014), reward signals are found in most midbrain dopamine neurons, as well as in subsets of neurons receiving dopaminergic projections in the orbitofrontal cortex (OFC), in the ventral striatum, and in the amygdala. These same neurons also respond to conditioned stimuli that serve to predict future reward. Other neurons connect information about reward with sensory or action information. Through associative reward learning, arbitrary(value) cues acquire the ability to automatically capture visual attention (Anderson, 2017). These are also neurons that discriminate sensory quality within and across categories. Rewards and reward expectation also affect the activity of more dorsal and medial regions of the prefrontal cortex that are involved in action selection and planning. In fact, many cortical (prefrontal, cingulate, parietal, inferior-temporal cortex) and subcortical (striatum, amygdala, superior colliculus) structures involved in high-level sensory and/or motor integration participate in the calculation of reward probability (Schultz, 2015). By extension, these rather complex sets of networks are involved in the related functions of attention, motivation, and acquisition. These networks are involved in information processing deficiencies associated with numerous emotional regulation disorders (Arsten & Rubia, 2012).

The brain does not simply operate to detect rewarding potential events or stimuli. Nonrewarding and aversive events are also represented in separate neuronal subpopulations in the midbrain, cortex, and amygdala. This is done with some degree of specificity as well. The identification of non-rewarding or aversive stimuli is an important component for the understanding of the connection between reward identification and attention. At the most basic level, it is equally important for an organism to know what stimuli could be effectively ignored and not attended to. This allows the individual to conserve cognitive and emotional resources. Additionally, it is important to know which stimuli have no intrinsic positive value but need to be attended to because they represent a threat to the individual. As an individual's primary function is to survive, they must be able to attend to potential threats in their environment. Going back to our lost dog, it would be very important to conduct an assessment in order to identify which dogs at the dog park might be aggressive. Obviously, this valuation would include an emotionally based component, perhaps heightened attention secondary to the observation of threatening behaviors. Avoiding these dogs would be a critical element of the search strategy.

There are, of course, networks that are devoted to the avoidance of aversive stimuli. Studies by Leathers and Olson (2012) suggest that cells in the posterior parietal cortex have been shown to encode both rewarding and aversive stimuli, which is seen as reflecting motivational salience rather than value (Leathers & Olson, 2012). These same researchers (Leathers & Olsen, 2017) later found that neurons in the lateral intraparietal (LIP) area of the parietal cortex respond to cues predicting rewards and penalties of variable size in a manner that depends on the motivational salience of the predicted outcome (strong for both large reward and large penalty) rather than on its value (positive for large reward and negative for large penalty). They considered this finding to suggest that the LIP mediates the capture of attention by salient events and does not encode value in the service of value-based decision-making. Their results indicated that neurons within the amygdala encoded cue value under circumstances in which LIP neurons exhibited sensitivity to motivational salience. As we have emphasized elsewhere (Wasserman & Wasserman, 2016), the inclusion of the amygdala indicates the inclusion of emotion-based factors in the determination of value. This conclusion has been supported by research that demonstrates that neurons located in the amygdala facilitate targeting eye movements (saccades) reliably based on aspects of emotional context, as is necessary for goal-directed and social behavior (Maeda et al., 2018).

Expectancy Theory

Human behavior, inclusive of attention, is generally guided by the anticipation of potential outcomes that are considered to be rewarding (Rademacher et al., 2010). Not all outcomes, however, are equally rewarding. As we have all experienced, some outcomes are more rewarding, some less so, and some not rewarding at all. Some outcomes are actually outright undesirable. In many cases the presentation of a stimulus is novel and therefore garners attention. These general concepts have been embodied in the work around Expectancy Theory (Eccles et al., 1983).

Expectancy Theory posits that low-level stimulus control of attention has been distinguished from high-level cognitive control of attention. Stimulus-driven attention refers to a transient bottom-up involuntary process. Attentional capture exemplifies this sort of low-level attentional control because it typically involves an immediate reaction to a salient or sudden and unexpected stimulus change (novel) that captures attention. High-level cognitive control of attention has been associated with voluntary top-down executive processes that guide attending toward a goal over the long term (Barnes & Jones, 2000).

Expectancy Theory then focuses on the controlling aspects of top-down and bottom-up systems on the regulation of attention. As far as this relates to what we are talking about here, this has implications for problems with the regulation of attentional processes that may appear in clinical practice. For example, individuals who do not find things to be novel, i.e., who do not recognize a salient feature, might be expected to have greater difficulty bringing attention to bear. An extension of this idea is demonstrated in work on post-error cognitive slowing (PES) in children with ADHD. This work shows reduced PES among children with an ADHD diagnosis is due to impaired error recognition during cognitively demanding tasks, defined as tasks that have a high cognitive load coupled with low error-cuing, consistent with impaired error recognition (Arnett et al., 2021). For certain individuals, many of whom have ADHD, errors are not novel and therefore not attention grabbing. There are no problems with the attention system directly but there are issues with its support systems. The result is that certain individuals will not self-correct errors and therefore appears that they are not "paying attention."

When applied to motivation, the Expectancy Theory of Motivation posits that people are motivated to do something because they anticipate their actions will lead to a highly desired outcome (Estes & Polnick, 2012). Essentially, Expectancy Theory hypothesizes that motivation is dependent upon the perceived association between performance of specific actions and their perceived possibility of obtaining certain outcomes. Part of this theory outlines how individuals modify their behavior based on their calculation regarding the probability that certain actions will achieve specific outcomes. Of course, in order for this to work, the individual needs to scan the environment and identify available rewarding stimuli. This is directly related to attention. Each stimulus in the environment has to be identified and the proper expectancy values assigned. Therefore, the precursor to the motivational activity is attention to the available stimuli. This attention is purposeful in that it is a prerequisite for value computations to be made. Once the best choice is identified, the individual needs to direct focus and relate effort to obtaining the stimulus. Obviously, problems can arise anywhere within the connectome related to this process.

The Relationship of Motivation, Anticipation of Reward, and Attention

One way to understand the concept of motivation from a neural network perspective is to apply the idea of anticipation of reward as the basis of its definition (Wasserman & Wasserman, 2020). There are neural network analogs that can be used to represent the intensity of the anticipation associated with a specific outcome and help us understand how the power of the anticipation of a particular outcome (set of stimuli) drives goal-directed behavior. Briefly, when competing choices are available, what you select to do is what you anticipate as bringing the highest value reward at the lowest expenditure of effort. Motivation is the summed balance between these two forces. Attention can be thought of in the same manner. The line between when an object is attention grabbing and then becomes motivating is blurry at best. These two constructs are clearly related as they are both governed by many of the same variables, especially perceived value. In many ways, what makes something attractive to attend to is the same thing that makes it motivating for the application of effort. The analog that most network models use to reflect these anticipatory choices is probabilistic reward calculations.

Attention plays a critical role in this process as it is the mechanism by which realistic choices are first identified in the environment. Attention is the mechanism that determines how much time is given over to the cognitive effort associated with considering each choice. Some choices may have value (Expectancy Value) but the effort to obtain them may, upon inspection, present too much cost (effort), offsetting motivation and winding up being discarded (no longer attended to).

As we have labored to point out motivation, affect, cognition, and attention are interrelated and may not be, on a functional level, completely dissociable (Drigas &

Karyotaki, 2019). Although reward values and emotion both affect attention, it is not clear whether their impacts are independent. For example, when a cue-target paradigm was used, behavior and eye tracking data were recorded in an emotion or sex recognition task under the conditions of reward and non-reward anticipation; the results showed that when a target was related to the emotional attribute of the stimulus, the reward promoted the processing target information, thereby generating reward-oriented attention (Yao et al., 2020). Positive reward and emotional valence are correlated with increased attentional focus. This brings up a number of interesting ideas, not the least of which is the idea that almost everything an individual encounters in the environment has both anticipatory and emotional valence. Therefore, is it reasonable to think that attention has properties of a state as well as properties of a trait? We think that it is.

Probabilistic Reward Calculations and the Human Brain

As we have discussed, people select targets to attend to and for potential action based upon the anticipation and multifactorial analysis of both rewards and punishments. One way to describe or understand the anticipatory process is to define it in a manner consistent with probabilistic reward theory. Understanding how a choice reflects selection properties of which choice is the most attention grabbing (salient) and attractive (motivating) is fairly simple and predictable when the options for response differ on only one dimension. For example, if people are offered a choice between two rewards that differ only in amount, they generally choose and focus on the larger, rather than the smaller, reward. If offered a choice between two rewards that differ only in waiting time for acquisition, people will likely choose the reward available sooner rather than the one available later, and if offered a choice between two options that differ only in probability of success, they tend to choose the option that they believe will be more likely available. A complementary set of principles for negative outcomes (e.g., smaller punishments will be chosen over larger ones) also applies. These behavioral tendencies obviously make both economic and evolutionary sense (Green & Myerson, 2004).

Things become more interesting and complex when choice options differ on more than one dimension, and that is where probabilistic reward modeling comes in. For example, what happens when a person has to choose between a smaller reward that is available immediately and a larger reward that is available at a later date? How is the relationship between these variables impacted when the larger, delayed reward is more certain than the smaller, immediately available reward? Does the magnitude of the difference impact the choice? For example, Billy goes grocery shopping for his next-door neighbor, Mr. Johnson. Mr. Johnson is short of cash and needs to go to the bank, but the bank is closed and he has to wait for 2 days. He tells Billy that he can give him 5 dollars for his effort today, but if he waits until Monday, he can give him 6 dollars. What does Billy do? Does this calculation change if the sum of 10 dollars is available? Does this change if Mr. Johnson is the forgetful type and there is a 50% chance that he will forget the discussion by Monday? Probabilistic reward modeling treats this series of choices as sort of a multiple regression equation, with each variable receiving a certain weighting (positive or negative). These equations result in one number that represents the strength of the choice; One number that represents how "motivating" that choice is in relation to the other available choices. Issues related to poor reward discounting are evident in individuals with ADHD (Costa-Dias et al., 2013). Let us take a moment to see what that might actually look like in ADHD.

It is becoming more recognized that ADHD affects or is the result of alterations in multiple circuits including cognitive (Sonuga-Barke, 2005) and extending to emotions (Roberts, 2023). In an interesting study investigating the functional connectivity of the nucleus accumbens, an area considered of significant importance in reward processing, in adolescents with ADHD, Costa-Dias et al. (2013) found atypical connectivity between the nucleus accumbens and several cortical and subcortical areas. They concluded that ADHD-related increased connectivity between the nucleus accumbens and the prefrontal cortex was associated with greater impulsivity secondary to steeper delayed-reward discounting. They further posit that these findings are consistent with the hypothesis that atypical signaling of the nucleus accumbens to the prefrontal cortex in persons with ADHD may lead to failure in estimating future consequences and less considered approach, i.e., impulsive behavior. This certainly holds importance for clinical understanding of how some individuals with ADHD present with reduced motivation and poor problem-solving strategies and often negative consequences in school, work, and/or and daily living situations.

Returning to understanding the more neurotypical process, there is strong behavioral and physiological evidence that the brain both represents probability distributions and performs probabilistic inferences on a continuous basis (Pouget et al., 2013). People perform near-optimal probabilistic inference in a wide range of cognitive and physical tasks. This probabilistic inference requires trial-to-trial representation of the uncertainties associated with task variables, retention, and subsequent use of this representation (Orhan & Wa, 2017). Essentially, these models use probability to describe how the human brain deals with uncertainty. The models are used to describe how the brain selects a course of action from among a group of actions where the known outcome is, in some ways, uncertain. Humans are continually presented with scenarios requiring assessment and decision-making toward a course of action. For example, should I watch TV now, call my friend, or do my homework? Should I do homework now, or can I wait to do my homework over the weekend? How important is it to do my homework? If I want to watch television, which television show do I wish to select? Do I start a new one or watch the remaining season of the one series I was watching? Is this a show my friend would be interested in? Do I care? How much?

Choosing (Choice Behavior) involves the weighing of multiple decision variables, such as usefulness, uncertainty, the amount of time between the behavior and the expected reward, or effort. All of these factors combine to create a subjective, relative value for each considered option, or potential course of action. This relative value also takes into account prior learning about potential rewards (and punishments) that result from prior actions. In a social context, decisions can also involve strategic thinking about the perceived intentions of others and about the impact of others' behavior on one's own outcome. Valuation is also influenced by different emotions that serve to adaptively regulate choices (Sirigu & Duhamel, 2016). All of the factors are considered one right after the other on an almost continuous basis. Not all human share the same optimal strategy. Some optimize reward rate while others emphasize accuracy. There is a range between these two poles. Research suggests that a substantial subset of individuals select decision thresholds that maximize reward rate, while the remainder select for higher thresholds of accuracy. Therefore, humans optimize a varying set of combination of reward rate and accuracy while at the same time, compensating for the influence of timing uncertainty, or both (Bogacz et al., 2010). Obviously, attentional skills such as focus and sustaining would be reflective of the strategy chosen.

Expected Value

One way to understand or investigate anticipation, as well as related attention and motivation, is to understand the concept of expected value (EV). In probabilistic decision-based modeling, the expected value of a particular choice is calculated, and after the choice has been made, the value of the choice can be updated based upon a temporal difference (TD) prediction error between the EV and the reward magnitude (RM) (actual reward) obtained. The EV is measured as the probability of obtaining a reward multiplied by the reward value placed on having the reward (Rolls et al., 2008). The choice and their respective EVs are retained in memory and serve as stimuli to drive future attention and subsequent goal-directed behavior.

Phases of Reward Calculation and Attention

As we have mentioned, reward calculation can be conceptualized as being comprised of three phases. The first phase is one of reward anticipation, which is the motivation to obtain and desire a reward, and the second phase is reward attainment, which describes the acquisition of the reward. The third phase is reward consumption, i.e., the liking and utilizing of the reward. Attention is informed by the results of all three phases in that top-down attention is modified by the valuations stored in memory.

With respect to reward calculation, people make choices in anticipation of rewards all the time. The choices are mediated by networks (cortico–basal ganglia–thalamo-cortico loop) related to emotional responsivity and cognitive framing. An individual then selects a response based upon a number of variables, including what has worked for us in the past, what has the highest probability of being successful in obtaining what we want, and what we perceive to be rewarding based upon our history, personal and cultural values and ethics, or in other words, our associative learning history, incentive valence, and positively valenced emotions. Sometimes it may happen that adaptive responses that worked for us in the past are no longer useful strategies. We then have to recalculate all of the above to determine our motivation, or the effort we are willing to expend to move forward in acquiring the reward. Our reward calculations, as well as both the subsequent emotional and behavioral outcomes, have implications for both mental health in general and therapy efficacy in particular. This places attention within a different framework. In order to do the above steps, an individual must attend in some manner to the results of the calculations being done. Even if these calculations are done very quickly and automatically (multiple times per second), the results must be considered (attended to) and compared to calculations from other options. Arguably, this too is attention, although this type of attention is not typically described. Typically, attention was described as occurring in relation to one stimulus. The precursor step, attending to what to attend to, was rarely considered.

Credit Assignment Problem and a Potential Solution

The understanding that selecting the preferred choice from among a number of available or potentially available options, being how the human brain picks the action it will engage in, only gets us so far as regards both attention and related motivation. That is because, in part, the relationships between various options' valuations are usually clear only in retrospect. That is, the analysis had to be conducted, the selection action tried, and a determination of its effectiveness made. Much of what we call motivation, and much of what drives future attention, is actually a memory of a particular analysis of an event and its result, and the ability of that memory to drive goal-directed behavior in the future. This memory is triggered whenever the stimuli is related to appear in the stimulus field. The process is automatic. When something good happens, we want to know why it happened and how to make it happen again. We "look" for it in the environment and quickly, and automatically, identify and attend to it when it appears. This knowledge, or memory, often happens in the absence of the physical presence of future events themselves. For example, let us suppose that I am currently living in the city and have no use for a car. I am planning a move to a rural area and have to think about transportation. I begin to "attend" to cars because I am going to require on in my new lifestyle. I have owned a vehicle in the past and can remember both the pros and cons of ownership. All of the prior decisions and analyses that went into that initial decision have been condensed into a few more wide-ranging comprehensive memories that I use to drive goal selection (Luyckx et al., 2019).

In reality, many events may precede a reward. How do we know, or decide, which of them caused it? This is a crucial aspect of both motivation and attention, because these causal attributions in memory serve as triggers for all elements of goal-directed behavior including attention. They are all, also, drivers of motivation. For example, a third grade child goes to school on a particular day and her mother stops to talk to the teacher on the car pick-up line after the school. The teacher reports to the mother on a number of things that went well for her daughter in the school that day, one of which was that her daughter achieved an "A" on the spelling test. The mom, obviously thrilled at the report, says to the daughter on the way home, "Honey, I am so proud of you about you getting that A on the spelling test." The probabilistic value of a good grade on the spelling test has been altered by the mother's positive feedback and assigned a higher weight. When the next spelling test occurs, the child will have retained that memory and used it to drive performance on the new test. She is now more motivated than before to do well. We may assume here that this is a healthy relationship between parent and child. As such, the daughter is more likely to pick out and attend to the spelling test as a vehicle for her to earn the praise of her mother.

This expectation that reward will be possible is known as reward motivation (Murty & Adcock, 2014). Reward motivation hypothetically results in enhanced memory encoding for salient events, even if those events are not explicitly associated with reward. This is because overinclusive memory encoding should occur only during reward motivation, because any salient event encountered (attended to) during reward motivation is potentially predictive of reward outcomes. Recent research (Murty & Adcock, 2014) suggests that there are defined neural networks that support this process. Functional magnetic resonance imaging during states of high- or low-reward motivation demonstrated motivation amplified hippocampal activation to, and declarative memory for, expectancy violations. Expectancy violations occur when the actual outcome is differentiated in a meaningful way from the expected outcome. Connectivity of the ventral tegmental area (VTA) with medial prefrontal, ventrolateral prefrontal, and visual cortices preceded and predicted this increase in hippocampal sensitivity. These findings demonstrate that reward motivation can enhance hippocampus-dependent memory.

Equifinality, Multifinality, and Counterfinality

There are several models of how conditions and preferences impact strategies used in the allocation of attention. All of these represent strategies employed in relation to the allocation of attention used by an individual in order to select a certain stimulus in relationship to obtaining a goal. Equifinality is the principle that recognizes that a given end state can be reached by different means or mechanisms, whereas multifinality is basically the opposite, suggesting that similar conditions or mechanisms can lead to dissimilar outcomes (Nusslock & Alloy, 2017). The multifinality model also refers to a circumstance wherein a single means is seen as serving more than one goal, thus affording the attainment of several objectives via a single activity. The counter finality effect implies that the more a means is perceived as detrimental to an alternative goal, the more it is perceived as instrumental to its focal goal (Schumpe et al., 2018). All of these options require the elements of the circumstances where the stimulus is recognized and encountered to be attended to and choices considered and eliminated or accepted. Obviously, these same concepts are also useful when considering the utility of a network model in relation to the construct of motivation, which occurs when a specific goal or choice is selected for action.

The research on these concepts is based on the idea that much of what we understand motivation to be is cognitive (Kruglanski et al., 2015). To the extent that attention, in this circumstance, is a top-down process of selection for arousal, it is also part of the overall process. That is because values, goals, wishes, and desires are mentally represented, as are their presumed means of attainment. These internalized preferences regulate attention to the extent that they represent the value of a stimulus and govern allocation of attention in a value-driven model. The basic idea of motivation, as well as its precondition, attention, refers to the building of relations between specific activities and the desirable states of affairs they are presumed to bring about. In creating such constructions, people plan out the motivational space they intend to navigate, and chart the trajectory that offers the best way to meet their goals. This, in effect, creates a "mental map" that contains a representation of the various goals that are salient for people when they are faced with specific tasks or situations. These mental maps also contain the representations of possible means to the goals in question, as well as obstacles that may interfere with the attainment of these goals. The final product is a motivationally relevant mental map that consists of interlocking configurations of goals and means, creating a conceptual network individuals are contemplating and taking into account when they are making their choices and carrying out their actions (Schumpe et al., 2018). Obviously, these mental maps regulate attention in that they essentially represent predetermined value that would direct attentional resources to certain stimuli in the environmental field.

The multifinality configuration refers to a situation wherein a single means is seen as serving more than one goal, thus affording the attainment of several objectives via a single activity. This circumstance would signal very high value and increases the likelihood of selection for the allocation of attention. The multifinality configuration maximizes value that accumulates across the different goals.

In contrast to the multifinality configuration, the idea of equifinality depicts a situation where in each of a different means leads to a single goal. This implies the need to exercise choice between the different means; it also extends the possibility of substituting, if need be, each of the means for any of the others. This process is related to error analysis in that the individual attends to the fact that the action failed to produce the goal and another course of action would be required.

Finally, the counterfinality configuration posits a pattern of goals–means relations wherein a means is seen to serve a given focal goal, which is at the same time believed to undermine another goal. This suggests that the selection of certain activities will enhance the value of a specific goal, while at the same time decreasing the value of several others. This phenomenon would come into play when considering a value-based model for attention allocation.

What is important to remember is that each of these models is based on the determination of value and the attachment of value to a particular goal or objective. The strength or amount of value placed on a particular outcome would determine the power of that goal to drive behavior and the willingness of the individual to engage in any of these strategies to obtain the outcome. All of these strategies are based on the idea that one goal supersedes all others in perceived value. Once the outcome is selected, it is conceivable that all, or a combination of these strategies, will be employed. For example, the individual might try a preferred means of obtaining a given goal and then failing that might default to less frequently preferred strategies. In any event, it is understandable that the choice of one goal decreases the attractiveness of competing goals and decreases their ability to drive behavior. In any of these circumstances, the choices are driven by the perceived relative value of the goals at hand.

Equifinality, in particular, has been associated with individuals with ADHD. Individuals with ADHD show heightened equifinality, related specifically to a hypodopaminergic allele. This implies heightened reward sensitivity and overarousal, thereby impeding rapidity of a specific response (Bunford et al., 2023).

Interaction of Emotions in the Operation of the Reward Recognition Network

Individuals do not calculate utilities (reward probabilities) explicitly or exactly. Each person has a unique set of values associated with stimuli and unique reinforcement maps. People tend to construct preferences based on their own experiences (Gottlieb et al., 2007) and the history of reward associated with them. People overweight, or underweight, small, medium-sized, and moderately large probabilities, and they also exaggerate risks. As we have seen there are also individuals who are overaroused by stimuli in general and this overarousal affects the relative values of the stimuli available. This suggests that people's experiences of events leak into decisions, even when risk information is explicitly provided (Kusev & van Schaik, 2011). The same can be said of attention. Reward experience and related valuation inform attentional allocation. As a result, choices for action depend strongly on emotionally filtered context, the type of options, the degree of affect associated with these options, and the nature of the presentation of the available options in the decision-making situation. In addition, both value and the probability of an outcome are assessed in a nonlinear fashion with higher values having decreased marginal gains, and losses being valued greater than gains. In general, low probabilities are overweighted and high probabilities are underweighted (Kahneman & Tversky, 1979). As we might speculate, emotions play a significant role in these weightings. As we have indicated, these weightings are stored in memory and inform and direct attentional allocation. Highly valued stimuli are highlighted in environmental searches, while lower valued stimuli are underweighted. Let us provide an example. Suppose you have decided to purchase a new car and you go to the dealership and "walk the lot" to see what is available. You have decided to purchase an SUV and what is more you have decided that you want a red one. In this model, in a top-down fashion, red and SUV are overweighted and the other attributes are underweighted. Your attention is both directed and narrowed by these decisions. While the current decisions are clearly top-down in nature, the history behind them is reflective of past experiences and rewards related to them. Why is red a color you prefer? Perhaps your favorite toy in infancy was red, and you retain positive, automatized emotions related to that color. At this point, the choice of color might reflect a long history of bottom-up, prelinguistic experiences.

There are models that extend the impact of emotionally regulated attention. These models have the affective regulation networks calculating their own probability weightings, and then interfacing with the reward network to produce a combined weighting of valuation. This dual system model (Mukherjee, 2010) incorporates (1) individual differences in disposition to rational versus emotional decision-making, (2) the affective nature of outcomes, and (3) different task constructs within its framework. As indicated, many of these emotional valuations have long experiential histories, making them somewhat intractable to modification. These predetermined attentional preferences can result in clinically disruptive behaviors.

Some Clinical Implications

The idea that emotional states in the present, as well as emotional states historically and experientially associated with certain stimuli and stored in memory, drive certain aspects of attention, has significant implications for the practicing clinician. In sum, the disruptions at attentional allocation and focus we traditionally attribute to disordered primary attentional systems may be, instead, attributable to poorly weighted value systems associated with past experiences from which the wrong conclusion has been drawn.

Attention Mood and Reward Expectation

Experiences affect mood, which in turn affects the interpretation and predictive weighting of future experiences. This relationship appears to be governed by two specific principles. First, mood depends on assessing how recent reward outcomes differ from expectations. Second, mood biases the way we attend to potential outcomes (e.g., rewards), and this bias affects learning about those outcomes. This two-way interaction serves to mitigate inefficiencies in the application of reinforcement learning to real-world problems. In this model, mood represents the overall momentum of recent outcomes, and its biasing influence on the perception of outcomes "corrects" learning to account for environmental dependencies. This results in potential dysfunction of adaptive behaviors that have the potential to contribute to the symptoms of mood disorders (Eldar et al., 2016).

Although other interventions need to be considered, for example psychopharmacological and or working on attention, we do recognize that a person must find value in the target of their efforts. This makes these reward processes critical targets for the therapeutic process when potentially dealing with attention, motivation, or mood disordered clients. The new, healthier choices we wish our clients to make are dependent on attending, learning, and automating the desirable outcomes of these selection processes. In addition, an individual's assessment of the success of the experiences they have while they are in treatment and actually attending to and attempting the adaptive behaviors and cognitions we suggest are crucial to the therapeutic process.

Core Brain Dimensions and Mental Health

The National Institute of Mental Health (NIMH) launched the Research Domain Criteria (RDoC) initiative (Insel, 2014) in an attempt to understand the nature of mental health and illness in terms of varying degrees of dysfunctions in general psychological/biological systems. RDoC is an "attempt to create a new kind of taxonomy for mental disorders by bringing the power of modern research approaches in genetics, neuroscience, and behavioral science to the problem of mental illness." As part of the criterion, the RDoC framework calls for "the study of empirically-derived fundamental dimensions as characterized by related behavioral/psychological and neurobiological data (e.g., reward valuation, working memory). RDoC also emphasizes full-range dimensional approaches (from typical to increasingly abnormal), neurodevelopment and environmental effects, and research designs that integrate data across behavioral, biological, and self-report measures" (Cuthburt, 2022). The inclusion of reward valuation implies that attention and its associated processes fall within the domain of the RDoC.

This initiative utilizes five major domains of human functioning. These domains are designed around emotion, cognition, motivation, and social behavior. Within each domain are behavioral elements, processes, mechanisms, and responses, called constructs that comprise different aspects of the overall range of functions. Of importance, (1) the constructs are studied along a span of functioning from normal to abnormal (2) with the understanding that each is part of an environmental and neurodevelopmental process, and therefore studied within that context. This initiative encourages the development of new ways of diagnosing disorders of mental health and neurobiologically based disorders based on core brain–behavior dimensions. Rather than work based on behavioral clusters/definitions, RDoC begins with our current understanding of brain–behavior dimensions and aims to link these dimensions to specific symptoms.

Consistent with the premise of our work, the RDoC identifies pathophysiological mechanisms that cut across, or are common to, multiple disorders of mental health. Identifying these mechanisms that underlie these trans-diagnostic symptom clusters can help break down the behaviorally and arbitrary distinctions between categorically defined mental health disorders and account for comorbidity among current DSM diagnostic categories (Nusslock & Alloy, 2017). In this context the interdependency of the constructs of motivation and attention becomes understandable as part of goal attainment just as attention and visual/auditory processing is related to survival. These processes rarely, in the operational world, stand alone and we propose that they are best understood within the context of everyday human functioning.

Gating

Gating describes neurological processes of filtering out redundant or unnecessary stimuli in the brain from all possible environmental stimuli. Essentially, gating is the process that modulates input into the brain and ensures that the system is not overloaded. This is a crucial management process of the brain because at any one point during the day, there is more information coming into the brain than can be effectively processed. Humans need a way to parse that information down to critical elements to which to attend. This parsing is not a passive process, because the human brain actively selects the information to which it will attend based upon a number of variables. Among these variables are prior knowledge, past history of reinforcement, and interest. Gating is a crucial aspect of attention (Liu et al., 2018).

Gating, which depends upon contribution of the reward system to the allocation of attention and working memory to specific stimuli, requires multiple brain systems operating in concert (McGinty et al., 2011). Learning, which occurs when certain stimuli are selected from the environment to be attended to, is not a passive acceptance of prepackaged knowledge that exists but involves learners (clients) actively engaging with the material and selecting from the material elements that are meaningful to them in a purposeful and directed way. What is gated, and what is not, is entirely dependent on the individual doing the gating, and the success in terms of having something gated is by no means predictable by those individuals seeking to impart the knowledge. Educators have long known this and have, whenever possible, designed their material to be as interesting as possible so that their charges will select it to gate. At least the good ones do. It is easy to observe gating in action. Many clients in our office complain that their children do not listen to them because they are focused on their video game consoles when they are being spoken to.

Role of Reward Recognition in the Gating Network

As we have seen, what gets attended to and then gated, or selected for action, is highly dependent on the history of reinforcement (valuation) associated with the action. In essence, the higher the probability of a perceived reinforcement, the more likely it is that the stimuli associated with that reinforcement will be gated to attention. In support of this fact, there is emerging research concerning the integration of the reward recognition network with the gating system. In the gating system, "reward is a central component for driving incentive-based learning, appropriate responses to stimuli, and the development of goal-directed behaviors" (Haber & Knutson, 2010, p. 4). It is also true that stimuli associated with negative outcomes are also gated. This is related to the perception of threat and the avoidance of non-survival.

There is substantial agreement concerning the cortical and subcortical structural network components for complex, goal-directed human behavior. Specifically, Koziol and Budding (2009) identified the sub-thalamic nucleus and ventral pallidum, the subiculum and related hippocampal areas, the lateral habenula, the mesopontine rostromedial tegmental nucleus, the extended amygdala, the bed nucleus of the stria terminals, and the hypothalamus as elements of a gating network. As is true of other networks, the reward network is not a fixed system. Koziol and Budding conclude by stating "one consistent point that became apparent was that brain regions cannot be simply labeled as either contributing, or not contributing, to behavior; rather, it's necessary to consider the specific circumstances under which the region is being engaged" (p. 356). Hart et al. (2014) pointed out that "considerable evidence suggests that distinct neural processes mediate the attention to, acquisition, and performance of goal-directed instrumental actions". They point to a cortical–dorsomedial striatal circuit as critical for the acquisition of goal-directed actions and a cortical–ventral striatal circuit appears as mediating instrumental performance. This essentially means that, as expected, automatized behaviors and emotional responses have separate components of the reward recognition network associated with them as these elements must be combined to produce complex behavior inclusive of attention.

There are other subcortical structures that play a significant role in both gating and reward recognition. One of these is the pedunculopontine nucleus (PPN) (also referred to as pedunculopontine tegmental nucleus, PPTN or PPTg), which is located in the brainstem to the rear of the substantia nigra and next to the superior cerebellar peduncle. The PPN is historically identified as one of the main components of the reticular-activating system (Garcia-Rill, 1991), which plays a major role in attention processing. The PPN projects to a wide variety of cortical and subcortical systems. The PPN plays a significant role in gating both sensorimotor and reward-related behavior (Diederich & Koch, 2005). Similarly, the nucleus accumbens (NAcc) has been identified as critical in the control of goal-directed behavior. Taha and Fields (2006) found that a subset of NAcc neurons demonstrated a long-lasting inhibition in firing rate, whose onset preceded initiation of goal-directed sequences of behavior and terminates at the conclusion of the sequence. This action, which precedes behavioral initiation, clearly indicates the allocation of attentional resources. This firing pattern suggested that, when active, these neurons inhibited goal-directed behaviors that permitted the gating of target behaviors.

Other cortical structures such as the caudate nucleus are active when learning relationships between stimuli and responses or categories and suggestive of the role of attention. For example, Seger and Cinotta (2005) found that activity associated with successful learning was localized in the body and tail of the caudate and putamen. Hippocampal activity was associated with receiving positive feedback, but not with correct classification. Successful learning correlated positively with activity in the body and tail of the caudate nucleus and negatively with activity in the hippocampus.

Motivation, Attention, and Gating

Gating and its related properties are neural networks that perform a critical role in motivation, and attention is a crucial component of gating. While most behavioral learning models conceptualize that reinforcement makes a specific action more likely, by causing it to be attended to and ultimately gated, the neural network model of both attention and motivation emphasizes that reward (value) makes a response to a certain stimulus more likely (Werner, 1994). While gating represents the process around the selection of actions, motivation represents the analysis and culmination (summary) of all the probabilistic reward valuations related to a specific goal. Motivation represents a state of readiness, or action potential, that is stored in long-term memory based upon these summaries. In the case of a newly experienced stimulus, pattern matching allows for these potentials, which have been computed before in relation to similar task to be associated with the new task. These action potentials have been engaged so many times that in relation to specific tasks, they become highly automated. In order for these action potentials to be initiated, they must be perceived as associated with specific stimuli in the environment. In other words, they must be attended to. They must be identified as existing in the environment and available to be acted upon.

This gating-based approach is based on the use of mental representations of reward calculations, stored in long-term memory, that estimate the distributions of probabilities generated by observing certain events, in particular contexts. There are obviously many types of such events that can be grouped together in schemata-based classifications. This would allow for certain classes of events (academics, sports games) that contain specific stimuli to be considered equally motivating. The probabilities that are measured are typically either the probability of observing a certain state, or the probability of observing particular transitions between states, or the probability of observing a particular state after having observed a given state (Oudeyer & Kaplan, 2009). The automated store representations would represent the potential for selection of a course of action. This stored selection bias would represent intrinsic motivation. These selection biases exist in a state of readiness. They are "activated" when it is perceived that they are available. This perception of availability represents the role of attention in this process. It is a reciprocal relationship between the stored representations and the visual or auditory recognition (attention) that they are available.

It is not clear from the extant literature whether these representations would represent only an analysis of value or would include the other critical variable to the calculation of motivation that is the analysis of the probability that the action would be successful in obtaining the goal. This second major factor in intrinsic motivation is based on assessment of competence that an individual has for achieving self-determined results or goals. Included herein is the concept of difficulty associated with obtaining the selected outcome. This would of course involve attention to details indicative of the probability of success. These details would be idiosyncratic to the individual. They might be related to the properties of the achievement process, rather than the meaning of the particular goal being achieved (Oudeyer & Kaplan, 2009). It is unclear at the current time if these mental representations are combined or are stored separately from value considerations, so it is unclear whether attention comes into play in more than one way. It is likely that both possibilities are present. While in some cases, competence-based and knowledge-based models of intrinsic motivation might be somewhat equivalent in the determination of the selection of outcome, they often produce very different behaviors. For the most part, the capacity to predict what happens in a situation is only weakly associated to the capacity to impact a situation in order to achieve a given self-determined goal.

For example, let us assume that Bill likes a specific type of car and has always wanted to buy one. Let us also assume that Bill has decided that the car would absorb too much of his income. It would be too difficult to maintain. Having thought about this many times, it is difficult to see how both value and probability of success would not be automated. Now let us assume that Bill gets a promotion and a raise. Many of the obstacles no longer exist. The probability of obtaining equation has to be adjusted. This situation would then represent a circumstance where the value calculation (I still want this) would occur before a new and separate probability of obtaining calculation. The differences between the old and new calculation would amount to a prediction error and would go on to regulate attention so that a new valuation, reflective of the new circumstance, would be created.

Attention-Gated Reinforcement Learning (AGREL)

We hope we have, at this point, clarified that motivation and attention are inexorably intertwined. This idea has been buttressed by work that attempts to integrate probabilistic reinforcement and the gating of attention as it relates to choice and goal directedness. The attention-gated reinforcement learning model is one of these attempts. This model incorporates two factors that modify synaptic plasticity and describes a process that leads to a response being gated:

1. a reinforcement signal that is homogeneous across the network and depends on the amount of reward obtained after a trial, and
2. an attentional feedback signal from the output layer that limits plasticity to those units at earlier processing levels that are crucial for the stimulus–response mapping.

The first factor is a global reward-related signal. This signal reaches all relevant synapses and is presumably implemented in the brain by the release of neuromodulators. In AGREL, the first factor represents the difference between the amount of reward that is expected and the amount that is actually obtained.

The second factor is a site-specific effect due to the feedback of neuronal activity, which assigns credit to sensory neurons that play a critical role in the selection of an action. In AGREL, a brain site-specific factor is used to represent the assignment credit to hidden units that make up a network that is responsible for the selected/preferred action. This is accomplished by a feedback signal from the winning output unit to the units that made it win. In AGREL, the winning motor program feeds its activity back to all elements of the selected/recruited network. These connections between feedforward and feedback areas are reciprocal. As a result, if network components have strong feedforward, they usually also receive strong feedback from these same network elements. The consequence of this reciprocity is that neurons that provide the strongest excitation to a particular motor program also receive the strongest feedback and are then more likely to be utilized when a target goal (or highly similar circumstance) is identified.

As we have stated elsewhere, it is clear that prior learning is a foundational component in the gating of attention and that the value attributed to that learning gates attention (Wasserman & Wasserman, 2015). It is also clear that the history of success or failure of that directed attention is stored in working memory to be called upon in future circumstances. It is therefore possible to conceptualize these stored memories regarding the value and probability of success as the neural network property of motivation.

Attention, Motivation, and Reward Recognition Network

Those who are familiar with our work will recognize that we have written extensively about the role of the reward recognition network in the development of motivation (Wasserman & Wasserman, 2020). Talking about this same network as a key component of attention highlights the idea that there are a core group of networks that are recruited for many tasks. These networks are not always on. They are recruited in a task-dependent manner with each component playing a specific role (Zuo et al., 2018). The coordination of neurocognitive networks appears to arise from a dynamic regime that balances counteracting tendencies toward integration and segregation (Alderson et al., 2020).

Metastability

The theory of metastability refers to the human brain's ability to integrate several functional parts and to produce neural oscillations in a cooperative and coordinated manner, providing the basis for conscious activity (Penrose & liebowitz, 1979). The concept of metastability represents an important theoretical solution to the requirement that local areas operate independently yet also combine, behave synergistically, and operate in sync with each other to produce a complex outcome. Metastability is also important as an observable phenomenon, furnishing a dynamical explanation for how large-scale brain regions coordinate their activity in space and time to support cortical function (Tognoli & Kelso, 2013). The operation of these networks acting in unison has been represented in models that describe the firing of large clusters of neurons firing at the same time. "In the human brain large numbers of excitatory and inhibitory neurons are recurrently connected so that the numerous positive and negative inputs that each neuron receives cancel out on average. Neuronal firing is therefore driven by fluctuations in the input and resembles the irregular and asynchronous activity observed in cortical in vivo data." This has been termed winnerless competition. The balanced network model has been extended to accommodate clusters of strongly interconnected excitatory neurons in order to explain persistent activity in working memory-related tasks (Rost et al., 2018).

Metastability is high in a system that visits a range of different states over time (dynamic flexibility), whereas both highly ordered and highly disordered states are characterized by low metastability (dynamic stability) and high and low phase synchrony, respectively (Hovel et al., 2020). These mutually related dynamics can be interpreted at both the functional level, in terms of neural flexibility, and at the cognitive level, in terms of behavioral flexibility. A concrete example of a metastable dynamical system is the "winnerless competition." However, metastable phenomena may emerge from a variety of underlying mechanisms (where certain conditions are satisfied) and it is in this broader sense that we use the term. Empirical and theoretical accounts indicate that the brain derives its behavior from its identity as a complex dynamical system operating in the metastable regime of its coordination dynamics.

Based upon the above, it therefore would be appropriate to ask what the role of the reward recognition network in each of the activities that characterize metastability.

A growing body of evidence indicates that attention and motivation are intimately tied. Evidence for attention–motivation interactions is suggested by findings that stimuli carrying motivational significance preferentially engage attention, including stimuli with positive emotional valence (Englemann et al., 2009). In the model we are discussing, orientating attention is a precursor and a facilitator of motivation. Without attention, motivation does not happen. The idea of valuation is critical to attention and therefore a critical component of motivation. Let us take a look at the following example. From the moment I was able to drive, I wanted to own a vintage Austin Healy roadster (a longstanding, preestablished bias). Recently, I was in the Winwood section of Miami and saw a show of vintage cars with a sign that said, "Own the car of your dreams." I walk into the show room to speak to a salesperson and find out just how much my dream would cost me, but I have already decided that the cost would be too high. In other words, I am motivated to look but not to buy unless it was the price of the century. My motivation is to satisfy my curiosity, nothing further. The obvious discriminator is the amount of effort I would likely have to make in order to purchase the car (Herculean) and how sure I would be that the effort would pay off (not sure at all). After all, there are other factors that contribute to the decision such as the home renovation projects that constitute my wife's wish list. Simply stated, the direct overall calculation of value is insufficient to impel me to action. Value plays two roles here. The first is the initial predisposition, and the second is whether that same value is sufficient to impel me to a certain action as my other choices also have value. This is a clear example of complex human behavior being built upon combinations of basic behaviors (Harvey, 2019). Harvey describes complex cognitive performance in terms of domains of functioning that are hierarchical in nature, with the bottom referring to more basic sensory and perceptual processes and the top referring to elements of executive functioning and cognitive control. These domains are not independent of each other, and executive functioning exerts control over the utilization of more basic processes. By extension the same can be said for the various neural networks that support these cognitive activities.

If we accept that value is a key construct in the generation and maintenance of attention, then it is logical to think about the behaviorally defined conditions that would occur when this system did not function as it should. This has been recognized. "Attention dysfunction also occurs as a nonspecific effect of neurophysiological factors that affect arousal and metabolic state. The components of attentional control are complex and are mediated by an elaborate and highly integrated series of neural systems" (Cohen et al., 2014). Unfortunately, the vast majority of the existing research identifies individuals with ADHD and then goes on not to describe the various clinical and behavioral characteristic of that population. Although the populations meet the behavioral criteria for the designation, it is far from clear that the underlying neural networks that support these behaviors are uniform.

We can conjecture what individuals with difficulty in managing value-based decisions might look like in terms of attentional deficiency. Attentional capture by valuable but task-irrelevant stimuli could direct behavior to unconventional and perhaps clinically undesirable behaviors. "Attentional capture by valuable but task-irrelevant stimuli could also confer adaptive advantages in many circumstances, leading the perceiver to orient to inconspicuous and/or unexpected reward-related stimuli. At the same time, however, attentional capture by reward-related stimuli (e.g., drugs of abuse, excessive food, or even irrelevant but rewarding information like an e-mail chime) can be maladaptive when it conflicts with contextually appropriate goals (e.g., intended abstinence from a drug or food)" (Andersen et al., 2011). Clearly, the treatment for such disorders would be quite different from targeting impulse control or executive regulation, which is currently used to treat many attentional problems.

References

Alderson, T., Bokde, A., Kelso, J., Mcquire, L., & Coyle, D. (2020, August 15). Metastable neural dynamics underlies cognitive performance across multiple behavioural paradigms. *Human Brain Mapping, 41*(12), 3212–3234. https://doi.org/10.1002/hbm.25009

Andersen, B., Lauremt, P., & Yantis, S. (2011). Value-driven attentional capture. *PNAS Biological Science, 108*(25), 10367–10371. https://doi.org/10.1073/pnas.1104047108

Anderson, B. (2017, March). Reward processing in the value-driven attention network: Reward signals tracking cue identity and location. *Social Cognitive and Affective Neuroscience, 12*(3), 461–467. https://doi.org/10.1093/scan/nsw141

Andoh, J., Matsushita, R., & Zatorre, R. J. (2015). Asymmetric interhemispheric transfer in the auditory network: Evidence from TMS, resting-state fMRI, and diffusion imaging. *Journal of Neuroscience, 35*(43), 14602–14611.

Andrade, M. N. (2023, December 14). *History of attention*. Retrieved from LibreTexts Social Sciences: https://socialsci.libretexts.org/Bookshelves/Psychology/Cognitive_Psychology/Cognitive_Psychology_

Armel, K., Beaumel, A., & Rangel, A. (2006). Biasing simple choices by manipulating relative visual attention. *Judgment and Decision Making, 3*(5), 396–403.

Arnett, A., Rhodes, C., & Rutter, T. (2021). Reduced error recognition explains post-error slowing differences among children with attention deficit hyperactivity disorder. *Journal of the International Neuropsychological Society, 28*(8), 810. https://doi.org/10.1017/S1355617721001065

Arsten, A., & Rubia, K. (2012). Neurobiological circuits regulating attention, cognitive control, motivation, and emotion: Disruptions in neurodevelopmental psychiatric disorders. *Journal of the American Academy of Child and Adolescent Psychiatry, 51*(4), 356–367. https://doi.org/10.1016/j.jaac.2012.01.008

Awh, E., Belopolsky, A., & Theeuwes, J. (2012, August). Top-down versus bottom-up attentional control: A failed theoretical dichotomy. *Trends in Cognitive Sciences, 16*(8), 437–443. https://doi.org/10.1016/j.tics.2012.06.010

Balkenius, C., Förster, A., Johansson, B., & Thorsteinsdottir, V. (2008). Anticipation in attention. In G. Pezzulo, M. Butz, C. Castelfranchi, & R. Falcone (Eds.), *The challenge of anticipation* (Lecture notes in computer science) (Vol. 5225). Springer. https://doi.org/10.1007/978-3-540-87702-8_4

Barnes, R., & Jones, M. (2000, November). Expectancy, attention, and time. *Cognitive Psychology, 41*(3), 254–311. https://doi.org/10.1006/cogp.2000.0738

Beckmann, C. F. (2005). Investigations into resting-state connectivity using independent component analysis. *Philosophical Transactions of the Royal Society B: Biological Sciences, 360*(1457), 1001–1013.

Berridge, K. C., & Robinson, T. E. (1998). What is the role of dopamine in reward: hedonic impact, reward learning, or incentive salience? *Brain Research Reviews, 28*(3), 309–369.

Blum, K., Li-Chuan Chen, A., Braverman, E., Comings, D., Chen, T., & Arcuri, V. (2008). Attention-deficit-hyperactivity disorder and reward deficiency syndrome. *Neuropsychiatric Disease and Treatment, 4*(5), 893–918. https://doi.org/10.2147/ndt.s2627

Bogacz, R., Hu, P., & Cohen, J. (2010). Do humans produce the speed–accuracy trade-off that maximizes reward rate? *Quarterly Journal of Experimental Psychology, 63*(5), 863–891. https://doi.org/10.1080/17470210903091643

Bossong, M., & Kahn, R. (2016). The salience of reward. *JAMA Psychiatry, 73*(8), 777–778. https://doi.org/10.1001/jamapsychiatry.2016.1134

Bradley, K. A., Case, J. A., Freed, R. D., Stern, E. R., & Gabbay, V. (2017). Neural correlates of RDoC reward constructs in adolescents with diverse psychiatric symptoms: A Reward Flanker Task pilot study. *Journal of Affective Disorders, 216*, 36–45.

Bronco, P. (2020, February 1). Mapping language with resting-state functional magnetic resonance imaging: A study on the functional profile of the language network. *Human Brain Mapping, 41*(2), 545–560. https://doi.org/10.1002/hbm.24821

Bunford, N., Hamori, G., Nemoda, Z., Angyal, N., Fiath, R. U., & Paszthy, B. (2023, July). The domain-variant indirect association between electrophysiological response to reward and ADHD presentations is moderated by dopaminergic polymorphisms. *Comprehensive Psychiatry, 124*, 152389. https://doi.org/10.1016/j.comppsych.2023.152389

Cohen, R. A., Malloy, P., Jenkins, M., & Paul, R. (2014). Disorders of attention. In M. Parsons, T. Hammeke, & P. J. Snyder (Eds.), *Clinical neuropsychology: A pocket handbook for assessment* (pp. 463–497). American Psychological Association. https://doi.org/10.1037/14339-021

Costa-Dias, T., Wilson, V., Bathula, D., Iyer, S., Mills, K., Thurlow, B., & Stevens, C. (2013, January). Reward circuit connectivity relates to delay discounting in children with attention-deficit/hyperactivity disorder. *European Neuropsychopharmacology, 23*(1), 33–45. https://doi.org/10.1016/j.euroneuro.2012.10.015

Cuthburt, B. (2022, April). Research domain criteria (RDoC): Progress and potential. *Current Directions of Psychological Science, 31*(2), 107–114. https://doi.org/10.1177/09637214211051363

Damoiseaux, J. S., Rombouts, S. A., Barkhof, F., Scheltens, P., Stam, C. J., Smith, S. M., & Beckmann, C. F. (2006). Consistent restingstate networks across healthy subjects. *Proceedings of the National Academy of Sciences, 103*(37), 13848–13853.

Diederich, K., & Koch, M. (2005). Role of the pedunculopontine tegmental nucleus in sensorimotor gating and reward-related behavior in rats. *Psychopharmacology, 179*, 402–408.

Dong, G., Lin, X., & Potenza, N. (2015, March 3). Decreased functional connectivity in an executive control network is related to impaired executive function in Internet gaming disorder. *Progress in Neuro-Psychopharmacology and Biological Psychiatry, 57*, 76–85. https://doi.org/10.1016/j.pnpbp.2014.10.012

Drigas, A., & Karyotaki, M. (2019). Attention and its role: Theories and models. *International Journal of Emerging Technologies in Learning, 14*(12), 169–182.

Durston, S., van Belle, J., & de Zeeuw, P. (2011, June 15). Differentiating frontostriatal and fronto-cerebellar circuits in attention-deficit/hyperactivity disorder. *Biological Psychiatry, 69*(12), 1178–1184. https://doi.org/10.1016/j.biopsych.2010.07.037

Eccles, J. S., Adler, T. F., Futterman, R., Goff, S. B., Kaczala, C. M., Meece, J. L., & Midgley, C. (1983). Expectancies, values, and academic behaviors. In J. T. Spence (Ed.), *Achievement and achievement motivation* (pp. 75–146). San Francisco, CA: W. H. Freeman.

Eldar, E., Rutledge, R. B., Dolan, R. J., & Niv, Y. (2016). Mood as representation of momentum. *Trends in Cognitive Sciences, 20*(1), 15–24.

Englemann, J., Damaraju, E., Padmala, S., & Pessoa, L. (2009). Combined effects of attention and motivation on visual task performance: Transient and sustained motivational effects. *Frontiers in Human Neuroscience, 3*, 4. https://doi.org/10.3389/neuro.09.004.2009

Estes, B., & Polnick, B. (2012). Examining motivation theory in higher education: An expectancy theory analysis of tenured faculty productivity. *International Journal of Management, Business, and Administration, 15*(1), 13–19.

Failing, M., & Theeuwes, J. (2018). Selection history: How reward modulates selectivity of visual attention. *Psychonomic Bulletin & Review, 25*, 514–538. https://doi.org/10.3758/s13423-017-1380-y

Fan, F., Liao, L., Lei, T., Zhao, T., Xia, M., Men, W., et al. (2021). Development of the default-mode network during childhood and adolescence: A longitudinal resting-state fMRI study. *NeuroImage, 226*, 1. https://doi.org/10.1016/j.neuroimage.2020.117581

Garcia-Rill, E. (1991). The pedunculopontine nucleus. *Progress in Neurobiology, 36*(5), 363–389.

Gooding, D. C., Braun, J. G., & Studer, J. A. (2006). Attentional network task performance in patients with schizophrenia–spectrum disorders: evidence of a specific deficit. *Schizophrenia Research, 88*(1-3), 169–178.

Gottlieb, D. A., Weiss, T., & Chapman, G. B. (2007). The format in which uncertainty information is presented affects decision biases. *Psychological Science, 18*(3), 240–246.

Goulden, N., Khusnulina, A., Davis, N. J., Bracewell, R. M., Bokde, A. L., McNulty, J. P., & Mullins, P. G. (2014). The salience network is responsible for switching between the default mode network and the central executive network: replication from DCM. *Neuroimage, 99*, 180–190.

Green, L., & Myerson, J. (2004). A discounting framework for choice with delayed and probabilistic rewards. *Psychological Bulletin, 130*(5), 769.

Haber, S. N., & Knutson, B. (2010). The reward circuit: linking primate anatomy and human imaging. *Neuropsychopharmacology, 35*(1), 4–26.

Hart, G., Leung, B. K., & Balleine, B. W. (2014). Dorsal and ventral streams: the distinct role of striatal subregions in the acquisition and performance of goal-directed actions. *Neurobiology of Learning and Memory, 108*, 104–118.

Harvey, P. (2019). Domains of cognition and their assessment. *Dialogues in Clinical Neuroscience*, 227–237. https://doi.org/10.31887/DCNS.2019.21.3/pharvey

Hovel, H., Viol, A., Loske, P., Merfort, L., & Vuksanović, V. (2020). Synchronization in functional networks of the human brain. *Journal of Non Linear Science, 30*, 2259–2282. https://doi.org/10.1007/s00332-018-9505-7

Hyman, A. (2005). Addiction: A disease of learning and memory. *The American Journal of Psychiatry, 162*, 1414. https://doi.org/10.1176/appi.ajp.162.8.1414

Insel, T. (2014). The NIMH research domain criteria (RDoC) project: Precision medicine for psychiatry. *The American Journal of Psychiatry, 171*(4), 395–397. https://doi.org/10.1176/appi.ajp.2014.14020138

James, H. (1890). *The principles of psychology*. Holt.

Joel, D. W. (2000). The connections of the dopaminergic system with the striatum in rats and primates: An analysis with respect to the functional and compartmental organization of the striatum. *Neuroscience, 96*, 451–474.

Kahneman, D., & Tversky, A. (1979). Prospect Theory: AN analysis of decision under risk. *Econometrica, 47*, 263–291.

Kelley, A. E., & Berridge, K. C. (2002). The neuroscience of natural rewards: relevance to addictive drugs. *Journal of Neuroscience, 22*(9), 3306–3311.

Koziol, L. F., & Budding, D. E. (2009). *Subcortical structures and cognition: Implications for neuropsychological assessment*. Springer.

Kruglanski, A. W., Chernikova, M., Babush, M., Dugas, M., & Schumpe, B. M. (2015). The architecture of goal systems: Multifinality, equifinality, and counterfinality in means—end relations. In *Advances in motivation science* (Vol. 2, pp. 69–98). Elsevier.

Kusev, P., & Van Schaik, P. (2011). Preferences under risk: content-dependent behavior and psychological processing. *Frontiers in Psychology, 2*, 269.

Leathers, M. L., & Olson, C. R. (2012). In monkeys making value-based decisions, LIP neurons encode cue salience and not action value. *Science, 338*(6103), 132–135.

Leathers, M. L., & Olson, C. R. (2017). In monkeys making value-based decisions, amygdala neurons are sensitive to cue value as distinct from cue salience. *Journal of Neurophysiology, 117*(4), 1499–1511.

Liu, J. C., Liu, K., & Zhao, J. (2018). Event detection via gated multilingual attention mechanism. In *Thirty-second AAAI conference on artificial intelligence*. New Orleans, LA: Association for the Advancement of Artificial Intelligence.

Lux, V. (2018, December). Epigenetic programming effects of early life stress: A dual-activation hypothesis. *Current Genomics, 19*(8), 638–652. https://doi.org/10.2174/1389202919666180307151358

Luyckx, F., Nili, H., Spitzer, B., & Summerfield, C. (2019). Neural structure mapping in human probabilistic reward learning. *elife, 8*, e42816.

Maeda, K., Kunimatsu, J., & Hikosaka, O. (2018). Amygdala activity for the modulation of goal-directed behavior in emotional contexts. *PLoS Biology, 16*(6), e2005339.

McGinty, V. B., Hayden, B. Y., Heilbronner, S. R., Dumont, E. C., Graves, S. M., Mirrione, M. M., ... & Haber, S. (2011). Emerging, reemerging, and forgotten brain areas of the reward circuit: Notes from the 2010 Motivational Neural Networks conference. *Behavioural Brain Research, 225*(1), 348–357.

Meek, W., & Benson, A. (2002, February). Dissecting the brain's internal clock: How frontal–striatal circuitry keeps time and shifts attention. *Brain and Cognition, 48*(1), 195–211. https://doi.org/10.1006/brcg.2001.1313

Menon, V. (2011). Large-scale brain networks and psychopathology: a unifying triple network model. *Trends in Cognitive Sciences, 15*(10), 483–506.

Menon, V. (2015). *Brain mapping: An encyclopedic reference* (Vol. 2). Elsevier: Academic Press.

Mukherjee, K. (2010). A dual system model of preferences under risk. *Psychological Review, 117*(1), 243.

Murty, V., & Adcock, R. (2014). Enriched encoding: Reward motivation organizes cortical networks for hippocampal detection of unexpected events. *Cerebral Cortex, 24*(8), 2160–2168.

Nestler, E. J. (2014). *Epigenetic Mechanisms of Depression. JAMA Psychiatry, 71*(4), 454–456.

Nusslock, R., & Alloy, L. B. (2017). Reward processing and mood-related symptoms: An RDoC and translational neuroscience perspective. *Journal of Affective Disorders, 216*, 3–16.

Orhan, A., & Wa, W. (2017). Efficient probablistic inference in general neural networks trained with non-probabalistic feedback. Nature. *Communications, 8(1)*, 138.

Oudeyer, P. Y., & Kaplan, F. (2009). What is intrinsic motivation? A typology of computational approaches. *Frontiers in Neurorobotics, 1*, 6.

Pace-Schott, E. P. (2017). *Principles and practice of sleep medicine*. Elsevier.

Pavlov, I. (1927). *Conditioned reflexes: An investigation of the physiological activity of the cerebral cortex*. Oxford University Press.

Penrose, O., & Liebowitz, J. (1979). Toward a rigorous molecular theory of metastability. In E. Montroll & J. Leibowitz (Eds.), *Studies in statistical mechanics II, fluctuation phenomena* (pp. 319–322). North Holland Publishing Company.

Pochon, J. B., Levy, R., Fossati, P., Lehericy, S., Poline, J. B., Pillon, B., ... & Dubois, B. (2002). The neural system that bridges reward and cognition in humans: an fMRI study. *Proceedings of the National Academy of Sciences, 99*(8), 5669–5674.

Pouget, A., Beck, J. M., Ma, W. J., & Latham, P. E. (2013). Probabilistic brains: knowns and unknowns. *Nature Neuroscience, 16*(9), 1170–1178.

Rademacher, L., Krach, S., Kohls, G., Irmak, A., Gründer, G., & Spreckelmeyer, K. N. (2010). Dissociation of neural networks for anticipation and consumption of monetary and social rewards. *Neuroimage, 49*(4), 3276–3285.

Raiker, J. S., Friedman, L. M., Orban, S., Kofler, M. J., Sarver, D. E., & Rapport, M. D. (2019). Phonological working memory deficits in ADHD revisited: The role of lower level information-processing deficits in impaired working memory performance. *Journal of Attention Disorders, 23*(6), 570–583. https://doi.org/10.1177/1087054716686182

Ranganath, C. (2010). A unified framework for the functional organization of the medial temporal lobes and the phenomenology of episodic memory. *Hippocampus, 20*(11), 1263–1290.

Roberts, C. S. (2023). Impact and centrality of attention dysregulation on cognition, anxiety and low mood in adolescents. *Scientific Reports, 13*, 9106.

Rolls, E., McCabe, C., & Redoute, J. (2008). Expected value, reward outcome, and temporal difference error representations in a probabilistic decision task. *Cerebral Cortex, 18*(3), 652–663.

Rost, T., Deger, M., & Nawrot, M. (2018). Winnerless competition in clustered balanced networks: Inhibitory assemblies do the trick. *Biological Cybernetics, 112*(5283), 81. https://doi.org/10.1007/s00422-017-0737-7

Sali, A., Aderson, B., & Yantis, S. (2014). The role of reward prediction in the control of attention. *Journal of Experimental Psychology: Human Perception and Performance, 40*(4), 1654–1664. https://doi.org/10.1037/a0037267

Schultz, W. (2015). Neuronal reward and decision signals: from theories to data. *Physiological Reviews, 95*(3), 853–951.

Science Direct. (2020). *Precuneus*. Retrieved from Science Direct: https://www.sciencedirect.com/topics/neuroscience/precuneus#:~:text=The%20precuneus%20is%20a%20brain,and%20affective%20responses%20to%20pain

Schumpe, B. M., Bélanger, J. J., Dugas, M., Erb, H. P., & Kruglanski, A. W. (2018). Counterfinality: On the increased perceived instrumentality of means to a goal. *Frontiers in Psychology, 9*, 1052.

Seeley, W. (2019). The salience network: A neural system for perceiving and responding to homeostatic demands. *Journal of Neuroscience, 39*(50), 9878–9882. https://doi.org/10.1523/JNEUROSCI.1138-17.2019

Seger, C. A., & Cincotta, C. M. (2005). The roles of the caudate nucleus in human classification learning. *Journal of Neuroscience, 25*(11), 2941–2951.

Sirigu, A., & Duhamel, J. R. (2016). Reward and decision processes in the brains of humans and nonhuman primates. *Dialogues in Clinical Neuroscience, 18*(1), 45–53.

Smallwood, J., Bernhardt, B. C., Leech, R., Bzdok, D., Jefferies, E., & Margulies, D. S. (2021). The default mode network in cognition: a topographical perspective. *Nature Reviews Neuroscience, 22*(8), 503–513.

Sohn, H., Meirhaeghe, N., Rajalingham, R., & Jazayeri, M. (2021, March). A network perspective on sensorimotor learning. *Trends in Neurosciences, 44*(3), 170–181. https://doi.org/10.1016/j.tins.2020.11.007

Sonuga-Barke, E. (2005). Causal models of attention-deficit/hyperactivity disorder: From common simple deficits to multiple developmental pathways. *Biological Psychiatry, 57*, 1231–1238.

Taha, S. A., & Fields, H. L. (2006). Inhibitions of nucleus accumbens neurons encode a gating signal for reward-directed behavior. *Journal of Neuroscience, 26*(1), 217–222.

Theeuwes, J. (2018, May 14). Visual selection: Usually fast and automatic; seldom slow and volitional. *Journal of Cognition, 1*(1), 29. https://doi.org/10.5334/joc.13

Tognoli, E., & Kelso, J. (2013, January 8). The metastable brain. *Neuron, 81*, 35. https://doi.org/10.1016/j.neuron.2013.12.022

Wasserman, T., & Wasserman, L. (2013). Toward an integrated model of executive functioning in children. *Journal of Applied Neuropsychology: Child, 2*(2), 88–96. https://doi.org/10.1080/21622965.2013.748394

Wasserman, T., & Wasserman, L. (2015). The misnomer of attention deficit hyperactivity disorder. *Applied Neuropsychology: Child, 4*, 1–7. https://doi.org/10.1080/21622965.2015.1005487

Wasserman, T., & Wasserman, L. D. (2016). *Depathologizing psychopathology: The neuroscience of mental illness and its treatment*. Springer.

Wasserman, T., & Wasserman, L. D. (2017). *Neurocognitive learning therapy: Theory and practice*. Springer International Publishing.

Wasserman, T., & Wasserman, L. (2020). *Motivation, effort and the neural network model*. Springer Nature.

Wasserman, T., & Wasserman, L. D. (2023). *Apraxia: The Neural Network Model*. Springer.

Werner, G. M. (1994, July). Using second order neural connections for motivation of behavioral choices. In *the third international conference on simulation of adaptative behavior* (pp. 154–161).

Yao, Y., Xuan, Y., Wu, R., & Sand, B. (2020). Regulatory effects of reward anticipation and target on attention processing of emotional stimulation. *Frontiers on Psychology*. https://doi.org/10.3389/fpsyg.2020.01170

Yi, W., Mei, S., Zhang, M., & Zheng, Y. (2020). Decomposing the effort paradox in reward processing: Time matters. *Neuropsychologia, 137*, 107311.

Zang, Y., Li, Q., Wang, Z., Liu, X., & Zheng, Y. (2017, September). Temporal dynamics of reward anticipation in the human brain. *Biological Psychology, 128*, 89–97. https://doi.org/10.1016/j.biopsycho.2017.07.011

Zuo, N., Yang, Z., Li, J., & Jiang, T. (2018, September). Core networks and their reconfiguration patterns across cognitive loads. *Human Brain Mapping, 39*(9), 3546–3557. https://doi.org/10.1002/hbm.24193

Chapter 9
Salience

Whether an object captures our attention depends, in part, on its salience. Salience, in this regard, is partially defined as how one object stands out from the others in its environment or how different a stimulus is compared with its neighbors in the visual field being considered (Melloni et al., 2012). We are going to talk about salience and its ability to direct attention. From an evolutionary standpoint, attention to things that "jump out" may be the difference between surviving and not. However, in some cases, the salient process drives our attention to suboptimal choices in the environment. Some of these suboptimal choices are core behaviors associated with numerous psychiatric conditions. For example, in the case of OCD, salience might direct our attention to variables that are actually detrimental to our functioning. Although historically salience and attention have been viewed as distinct processes, when considering salience in the context of an integrated network model, it represents a necessary precondition to the allocation of attention.

How our attention is garnered based on how different something is is known as bottom-up salience. In essence, it catches our attention. How it compares based on internal directives is called top-down control. In essence, top-down control reflects our goals, values, and preferences.

As we develop, top-down salience may become more impactful. However, as we develop, these two disparate networks become highly integrated and non-dissociable on a functional level. Of course, this process usually requires both the "attention grab" and the "goal-oriented attention" in order to guide our behavior. In this chapter, we will focus, to the extent possible, on each component separately. Where possible, we will segregate these two integrated components for the purpose of attempting to understand what could go wrong with either element of salience that could disrupt attention. Most research regarding bottom-up attentional processes utilizes visual search paradigms. This is because, as language inevitably contains a top-down component, it would confabulate the data. Traditionally, visual search is divided into two elements: one where target selection is purely bottom-up (sensory)

T. Wasserman, L. D. Wasserman, *Disorders of Attention*, Neural Network Model: Applications and Implications, https://doi.org/10.1007/978-3-031-78732-4_9

driven and another that requires top-down control. In bottom-up processing, an object stands out and captures attention due to its saliency (e.g., one person wearing red when everyone else is wearing black). In the top-down processing, endogenous control is required to meet the observer's goals. These two forms of attention must be coordinated to guide behavior.

In both natural and laboratory settings, it has been demonstrated that almost no search is purely bottom-up or purely top-down driven. In the top-down procedure, high-level semantic knowledge or contextual information is believed to guide both information processing and the refinement of attentional focus. The bottom-up process, in contrast, is carried out in the opposite direction, from the stimulus to high-level visual understanding. This is not a one-time interaction. These processes interact in an iterative fashion, with each successive stage performing an ever more complex analysis and refining of the input (Wang et al., 2019).

One model of salience, in general (Menon & Uddin, 2010), points to the anterior insula as an integral hub in mediating dynamic interactions between other large-scale brain networks that are involved in externally oriented attention (bottom-up) and internally oriented or self-related cognition (top-down). The model hypothesizes that the insula is responsive to salient events and that its basic function is to identify these designated salient events for additional processing and initiate appropriate control mechanisms. They propose that the anterior insula and the anterior cingulate cortex form a "salience network" that functions to segregate the most relevant among internal and external stimuli in order to direct behavior. Within the framework, the various functions attributed to the insula can be conceptualized by a few basic mechanisms: "(1) bottom-up detection of salient events, (2) switching between other large-scale networks to facilitate access to attention and working memory resources when a salient event is detected, (3) interaction of the anterior and posterior insula to modulate autonomic reactivity to salient stimuli, and (4) strong functional coupling with the anterior cingulate cortex that facilitates rapid access to the motor system" (p. 655). In this manner, with the insula as its integral hub, the salience network aids target brain areas in the initiation of appropriate behavioral responses to salient stimuli.

Another model of salience seeks to segregate it as a separate but highly integrated component of the attention process (Parr & Friston, 2019). This theory would go along with our previous work suggesting that processes like executive function, working memory, and now attention represent initially separate functions that become non-dissociable, integrated as a person neurodevelopmentally progresses. In this model, attentional processes select those sensations that are informative about their causes. These processes can be separated into two categories: those that enhance the weighting of current data and those involved with the immediate acquisition of new sensations. While the former makes use of sensory data in the present, the latter is concerned with the future. This model, therefore, separates attention and salience.

Salience is associated with planned actions that resolve uncertainty, while attention is a process of (precision-dependent) gain control. In this model, attention is a consequence of the need to actively engage with the sensory world around us. Much of this model is based on the research on sensory gain control.

Sensory Gain Control

Gain control serves to reduce the weighting of undesired or disruptive sensory input while selectively amplifying feedback useful for adjusting ongoing movements and shaping motor output to match changing environmental demands. Both physiological and behavioral studies have demonstrated that stimulus-driven neural activity in various sensory pathways can be modified in amplitude during selective attention, especially during the earlier phases. Recordings of event-related brain potentials indicate that such sensory gain control or amplification processes play an important role in visual-spatial attention (Hillyard et al., 1998). Essentially, attended information elicits larger sensory-evoked responses than non-attended or ignored information. In the earlier stages of this process, selection is considered covert as no behavioral action is taken. The selected information is amplified and occurs in a wide variety of sensory situations. This early amplification represents saliency, or the highlighting of important information. The focus on the highlighted stimulus (action) represents attention.

A central concept in understanding the difference between the covert (salience) and overt (attention) sampling of our sensorium is based on a predictive model of attention. In this model, the brain possesses an internal model of the world around it that tries to account for how sensations are generated. Perception is a process of optimizing beliefs about the causes of sensations, while actions select new sensory data to test hypotheses about these causes (Friston et al., 2012). Connections from the visual cortex to the dorsal ("where") and ventral ("what") streams can be thought of as carrying evidence in favor of spatial configurations or stimulus identities. Those from frontal areas to posterior cortices may signal predictions about the "what" and "where" information that would be obtained as a consequence of action. On the other hand, they may modulate the gain of signals in these pathways. Salient actions are those that solicit unambiguous data that could not have been predicted with a high degree of certainty prior to that action. Based upon this line of research, salience has been defined in terms of intrinsic motivation, or Bayesian surprise (Itti & Baldi, 2006), among other labels. It cannot be stressed highly enough, that these distinctions, representing separate network processes, are mostly for the purposes of understanding the complexity of the processes. In functional reality, for anyone past initial infancy, these processes are steadily integrating and, by adulthood, are dissociable.

Given the focus of this book, the question should be asked about whether or not a disruption of salience as defined by this model results in disorders of attention. The answer is that there are. Gain control is central in balancing influences on perception from competing sensory evidence and from internally generated prior beliefs. Failures in doing so lead to false perceptions, including hallucinations. Syndromes of aberrant salience computation manifest as failures of active and accurate scene construction (Parr & Friston, 2019).

Salience Detection Network Components

Salience detection is a complex mechanism that requires a bottom-up mechanism for filtering stimuli standing out from a stream of sensory inputs and a higher-order (top-down) mechanism for the automatic attraction and consequent maintenance of attention on a specific task. The salience network (SN) plays a dominant role in the detection of salient stimuli across multiple modalities (Onofrj et al., 2022). As we have seen, the insula, together with the anterior cingulate cortex, constitutes the main anatomical structures to which the SN is anchored (Menon & Uddin, 2010). Saliency detection is also involved in the dynamic interaction of sensory and cognitive influences that control attention (Menon & Uddin, 2010). Attention is controlled by two partially segregated networks: the ventral attention network (VAN) and dorsal attention network (DAN), forming a twofold attentional control system (Vossel et al., 2014). The DAN shows sustained activation when focusing attention on an object. This complex interaction of bottom-up and top-down attention mechanisms also requires the ability to disengage from the constant stream of our self-referential thoughts referred to as "mind wondering," a cognitive process attributed to the DMN (Raichle, 2015; Smallwood et al., 2021). This hypothesis is supported by the research that performance of a goal-directed, non-self-referential task is accompanied by a decrease in activity in the DMN and a corresponding increase in activity in the DAN (Esposito et al., 2018).

Dorsal and Ventral Attention Networks

Many models of spatial attention focus on dissociable but interacting neural pathways in the brain's visual system as a fundamental component of the saliency system. Projecting from the primary visual cortex in the occipital lobe, two neural pathways arise. The ventral network (VAN) comprises the temporoparietal junction (TPJ) and the ventral frontal cortex (VFC) and typically responds when behaviorally relevant stimuli occur unexpectedly (e.g., when they appear outside the cued focus of spatial attention). It extends ventrally into the temporal lobe and appears to largely encode what is in the observed visual field, recognizing objects. It has been proposed that the ventral system is lateralized to the right hemisphere of the brain. Whereas functional imaging studies indeed more consistently report right-hemispheric activation in temporoparietal areas, the left TPJ has also been shown to subserve attentional functions, and several studies have observed bilateral TPJ activation in tasks tapping attentional reorienting and the processing of rare deviant stimuli. The VAN is generally activated when an unexpected event occurs, causing one's attention to divert from the current task (bottom-up processing).

The dorsal network is organized bilaterally and comprises the intraparietal sulcus (IPS) and the frontal eye fields (FEF) of each hemisphere. Passing dorsally into the parietal lobe, it provides information about where objects are located in the visual field. Looking at the way these pathways interact, the ventral network uses object classification information to guide attention to relevant image regions, and the dorsal network uses this attentional information to accurately localize and identify objects in the scene (Ebrahimpour et al., 2019). The DAN is thought to be responsible for goal-directed, top-down processing of attention.

Saliency Maps

One way to understand the allocation of attention is that of a saliency map. This "map" would pool together all types of feature information (e.g., shape, size, color, and orientation) in the visual field (Koch & Ullman, 1985). Attention is hypothetically allocated to the location with the highest activity on this global map. Candidate brain areas would have access to a wide array of visual information, therefore most likely being areas found at the later stages of visual cortical pathways. Considering that the generation of eye movements is intricately intertwined with the allocation of attention, the neural site of the saliency map is also likely to be closely related to eye movement preparation. Studies have suggested multiple brain areas that could serve as maps of saliency, including the frontal eye field (FEF) and area 46 of the prefrontal cortex (PFC), area 7a and the lateral intraparietal area (LIP) in the posterior parietal cortex (PCI), the superior colliculus, and the substantia nigra (Katsuki & Constantinidis, 2016). Anatomically, areas in the PFC and PPC such as the FEF and LIP have reciprocal projections to each other and direct connections with many other extrastriate areas in the dorsal and ventral pathways. The FEF and LIP also send output signals and receive indirect inputs from the superior colliculus. It has been well recognized that prefrontal areas such as the dlPFC and FEF represent salient stimuli in "oddball" stimulus tasks requiring detection or a response towards a stimulus differing in color or shape from background elements (Sato et al., 2003). Demonstrations of bottom-up saliency reflected in the activity of PFC and PPC neurons are not limited to oddball stimulus paradigms. The idea of these maps would be to summarize all activity related to the saliency of a particular object in a particular visual field. To date, they remain conceptual in human neural networks but are employed in computational modeling.

Disorders of Salience

In addition to the disorders identified above, there is research that identifies saliency deficits in attention functioning. For example, there is research that helps explain the poor allocation of attention in children presenting with the symptoms of ADHD. FMRI imaging in youths ages 11–16 years provides evidence that ADHD patients show enhanced neural activity in response to novel but behaviorally irrelevant stimuli, as well as showing reduced habituation to familiar items. These findings suggest an inefficient use of neuronal resources in children with ADHD that could be closely linked to increased distractibility and directly related to salience (Tegelbeckers et al., 2015). Interestingly, this study demonstrated that novel stimuli activated a comparable network of brain regions in children and adolescents with and without ADHD. This network included the bilateral parahippocampal and fusiform gyrus, the hippocampus, and the temporal, as well as occipital gyri. This network is consistent with findings of novelty processing areas in visual oddball tasks in healthy adults. The medial temporal areas and especially the hippocampus are areas known to detect deviance from expectation and novelty by matching incoming information with memory content. Increased activation in visual association cortices, such as the fusiform gyrus and occipital regions, has been associated with attention shifts driven by novel stimuli. In other words, the problem occurred in the identification of the designated salient stimuli before attention was allocated. That is, the salience-based distraction was not directly associated with the actual functioning of the classically defined attention networks, which were functioning correctly.

There is additional research that provides evidence that decreased functional connectivity within a single salience/ventral attention subnetwork, as well as between two frontoparietal subnetworks, significantly correlates with ADHD symptoms presented by Yerys et al. (2019).

Dysfunction involving the salience network (SN) is connected to a broad array of deficits and maladaptive behavioral patterns in several clinical disorders. These include depression, insomnia, narcissism, PTSD (in the case of SN hyperactivity), chronic pain, anxiety, high degrees of neuroticism, schizophrenia, epilepsy, autism, neurodegenerative illnesses, and bipolar disorder (in the case of SN hypoactivity). In general, these conditions can be associated with a widespread disruption in predictive coding at multiple hierarchical levels (Schimmelpfennig et al., 2023). This lack of effective predictive coding implies that attention, in these and perhaps other disorders, is disrupted and would account for the finding of co-morbidity of attention deficit disorder with the disorders indicated above.

Let's take a look at an example to see how this might look in clinical practice. Let's say that a soldier has just returned from a war zone and is manifesting signs of post-traumatic stress disorder and related depression. He/she is particularly stressed and disoriented by loud bangs in the environment. The soldier is now almost constantly scanning the environment for such occurrences, and as a result, his attention to his work and his marital relationship have been affected. The soldier cannot sustain attaining a task and is almost always distracted. He/she is always scanning the environment and frequently gets up in the middle of an activity to check on a noise in the environment. In short, the soldier now demonstrates the requisite criteria for ADHD as it is currently defined. Do you, in practice, diagnose the ADHD? The answer is, of course, that you do not because the criteria for ADHD include the caveat that the symptoms are not better explained by another mental disorder (such as a mood disorder, anxiety disorder, dissociative disorder, or personality disorder). The symptoms do not happen only during the course of schizophrenia or another psychotic disorder. Or that the individual did not display symptoms before the age of 12. Functionally, however, salience and related attention have been disrupted, and the soldier cannot now pay attention. This is the same disruption that many individuals with ADHD demonstrate. The only difference is that we hypothetically know the cause in the case of the soldier, and we do not yet know the cause in the case of ADHD. In ADHD, we assume that there is a white matter dysfunction in the attention networks, but as we have seen, that might not always be the case.

Saliency Overactivity

Research of the salience network (SN) indicates that anterior insula overactivation leads to excessive sensitivity and anxiety arousal and maybe a joint characteristic in numerous conditions. This is probably related to the low excitability threshold of the structures that make up the SN (particularly the right anterior insula) and the classification of excessive stimuli as important. This leads to the generalization and over-mobilization of stress reactions in non-threatening situations (Hermans et al., 2014). In the case of narcissism, on the other hand, a proposed model of SN dysfunction relies on the inability of a dysregulated right anterior insula to turn off the default mode network (DMN), leading to an excessive concentration of thoughts on oneself (Jankowiak-Siuda & Zajkowski, 2013).

It is worth mentioning that SN hyperactivity is linked to the perception of physical symptomology. For example, the volume of gray matter within the insula and ACC, among others, is characteristically high in patients suffering from chronic pain (Cauda et al., 2017), and the subjectively perceived level of pain is correlated with the strength of AI and ACC activations (Legrain et al., 2011).

Saliency Underactivity (Underperformance)

Problems with an underperforming saliency network have been linked to cognitive and affective deficits in disorders such as schizophrenia, autism, and bipolar disorder (Schimmelpfennig et al., 2023). In schizophrenia, the key SN nodes' incorporation of excitation-inhibition balance is impacted by the pathophysiology of glutamate neurotransmission (Limongi et al., 2020). As it relates to this book, research links schizophrenia symptoms to abnormal signaling of reward, value-based decision-making, lack of long-term stability of internal models, and prior memory (Sterzer et al., 2019). In schizophrenia, long-standing unpredictability about upcoming sensory inputs finally leads to stimulus avoidance and psychomotor poverty (Corlett et al., 2016). In summary, the symptoms of schizophrenia are consistent with a decrease in high-level precision or a failure of sensory attenuation (an overestimate of the trustworthiness of the stimuli) leading to false inferences and failure in cognitive control, as well as the possibility of hallucinations and delusions.

As pertains to the current discussion, it would be beneficial to remember that the function of salience remains the same in all of these disorders. What changes is the dysfunction within that system, resulting in stimuli being overattenuated or underattenuated.

In obsessive-compulsive disorder (OCD), certain stimuli become overattenuated, have disproportionately increased value, and become the individual's focus. Deficits in attention have in fact been identified in obsessive-compulsive disorder (OCD) (Posner et al., 2017). In OCD, functional connectivity between two neural networks associated with attentional processes—the default mode network (DMN) and the salience network (SN)—are altered. In OCD, there is significantly reduced inverse connectivity between the anterior medial prefrontal cortex (amPFC) and the anterior insular cortex, regions within the DMN and SN, respectively. In OCD, reduced inverse DMN-SN connectivity has been associated with both increased OCD symptom severity and decreased sustained attention. OCD symptoms have been identified with an imbalance in DMN-SN networks that subserve attentional processes, and this effect of OCD on DMN-SN connectivity is associated with decreased sustained attention. This research adds to a growing literature (Fitzgerald et al., 2010; Beucke et al., 2014) suggesting that reduced inverse DMN-SN connectivity may represent a trans-diagnostic marker of attentional processes and suggests a potential mechanistic account of the relationship between OCD and attention. This same deficiency may appear in other disorders and represents a totally discrete class of attentional disorders. Reduced inverse DMN-SN connectivity may be an important target for the development of treatments to improve attention in individuals with OCD that differ from those that address other types of attentional disorders. Additional research suggests that the central executive network also interacts with the DMN and SN networks to produce the functional attentional deficits found in OCD (Fan et al., 2021).

In depression, there is a diminution of the value of almost all stimuli, with the possible exception of the top-down negative attributions associated with the disorder. Major depressive disorder (MDD) is characterized by altered inherent functional connectivity within (intra-iFC) intrinsic connectivity networks (ICNs), such as the Default Mode- (DMN), Salience- (SN), and Central Executive Network (CEN). There is a model that suggests that unusual switching between DMN-mediated self-referential and CEN-mediated goal-directed cognitive processes contributes to MDD, possibly explaining patients' difficulties to disengage the processing of self-focused, often negatively biased thoughts (Manoliu et al., 2014).

In summary, there appears to be a group of salience-based deficits in attentional functioning that occur in multiple disorders of mental health. Some salience-based disorders appear to be distinct from those found in ADHD. These deficits would appear to be potential targets for intervention in their own right and may be responsive to different treatment options than those currently in use for the attentional issues related to ADHD.

References

Beucke, J., Sepulcre, J. E., Sebold, M., Kathmann, N., & Kaufmann, C. (2014). Default mode network subsystem alterations in obsessive-compulsive disorder. *British Journal of Psychiatry, 205*, 376–382. https://doi.org/10.1192/bjp.bp.113.137380

Cauda, F., Costa, T., Nani, A., Fava, L., Palermo, S., Bianco, F., ... & Keller, R. (2017). Are schizophrenia, autistic, and obsessive spectrum disorders dissociable on the basis of neuroimaging morphological findings?: A voxel-based meta-analysis. *Autism Research, 10*(6), 1079–1095.

Corlett, P., Honey, G., & Fletcher, P. (2016). Prediction error, ketamine and psychosis: An updated model. *Journal of Psychopharmacology, 30*, 1145–1155. https://doi.org/10.1177/0269881116650087

Ebrahimpour, M., et al. (2019). Ventral-dorsal neural networks: Object detection via selective attention. In *2019 IEEE Winter Conference on Applications of Computer Vision (WACV)* (pp. 986–994). IEEE. https://doi.org/10.1109/WACV.2019.00110

Esposito, R., Cieri, F., Chiacchiaretta, P., Cera, N., Lauriola, M., Di Giannantoni, M., et al. (2018). Modifications in resting state functional anticorrelation between default mode network and dorsal attention network: Comparison among young adults, healthy elders and mild cognitive impairment patients. *Brain Imaging and Behavior, 12*(1), 127–141. https://doi.org/10.1007/S11682-017-9686-Y

Fan, F., Liao, L., Lei, T., Zhao, T., Xia, M., Men, W., et al. (2021). Development of the default-mode network during childhood and adolescence: A longitudinal resting-state fMRI study. *NeuroImage, 226*, 1. https://doi.org/10.1016/j.neuroimage.2020.117581

Fitzgerald, K., Stern, E., Angstadt, M., Nicholson-Muth, K., Maynor, M. W. R., et al. (2010). Altered function and connectivity of the medial frontal cortex in pediatric obsessive-compulsive disorder. *Biological Psychiatry, 68*, 1039–1047. https://doi.org/10.1016/j.biopsych.2010.08.018

Friston, K., Adams, R., & P. L., & Breakspear, M. (2012). Perceptions as hypotheses: Saccades as experiments. *Frontiers in Psychology, 3*, 151.

Hermans, E., Henckens, M., & Joëls, M. G. (2014). Dynamic adaptation of large-scale brain networks in response to acute stressors. *Trends in Neuroscience, 37*, 304–314. https://doi.org/10.1016/j.tins.2014.03.006

Hillyard, S., Vogel, E., & Luck, S. (1998). Sensory gain control (amplification) as a mechanism of selective attention: Electrophysiological and neuroimaging evidence. *Philosophical Transactions of the Royal Sociaty of Londondon Biological Science, 353*(1373), 1257–1270. https://doi.org/10.1098/rstb.1998.0281

Itti, L., & Baldi, P. (2006). Bayesian surprise attracts human attention. *Advances in Neural Information Processing Systems, 18*, 547.

Jankowiak-Siuda, K., & Zajkowski, W. (2013). A neural model of mechanisms of empathy deficits in narcissism. *Medical Science Monitor: International Medical Journal of Experimental and Clinical Research, 19*, 934–941. https://doi.org/10.12659/MSM.889593

Katsuki, F., & Constantinidis, C. (2016). Bottom-up and top-down attention: Different processes and overlapping neural systems. *The Neuroscientist, XX*(X), 1–13. https://doi.org/10.1177/1073858413514136

Koch, C., & Ullman, S. (1985). Shifts in selective visual attention: Towards the underlying neural circuitry. *Human Neurobiology, 4*(2), 19–27.

Legrain, V., Plaghki, L., & Mouraux, A. (2011). The pain matrix reloaded. *Progress in Neurobiology, 93*, 111–124. https://doi.org/10.1016/j.pneurobio.2010.10.005

Limongi, R., Jeon, P., Mackinley, M., Das, T., Dempster, K., Théberge, J., Bartha, R., Wong, D., & Palaniyappan, L. (2020). Glutamate and dysconnection in the salience network: Neurochemical, effective connectivity and computational evidence in schizophrenia. *Biological Psychiatry, 88*, 273–281. https://doi.org/10.1016/j.biopsych.2020.01.021

Manoliu, A., Meng, C., Brandl, F., Doll, A., Tahmasian, M., Scherr, M., Schwerthöffer, D., Zimmer, C., Förstl, H., Bäuml, J., & Riedl, V. (2014). Insular dysfunction within the salience network is associated with severity of symptoms and aberrant inter-network connectivity in major depressive disorder. *Frontiers in Human Neuroscience, 7*, 930. https://doi.org/10.3389/fnhum.2013.00930

Melloni, L., van Leeuwen, S., Alink, A., & Muller, N. (2012). Interaction between bottom-up saliency and top-down control: How saliency maps are created in the human brain. *Cerebral Cortex, 22*(12), 2943–2052. https://doi.org/10.1093/cercor/bhr384

Menon, V., & Uddin, L. (2010). Saliency, switching, attention and control: A network model of insula function. *Brain Structure and Function, 214*(5–6), 655–667. https://doi.org/10.1007/s00429-010-0262-0

Onofrj, V., Chiarelli, A., Wise, R. C., & Caulo, M. (2022). Interaction of the salience network, ventral attention network, dorsal attention network and default mode network in neonates and early development of the bottom-up attention system. *Brain Structure and Function, 227*, 1843. https://doi.org/10.1007/s00429-022-02477-y

Parr, T., & Friston, K. (2019). Attention or salience? *Current Opinion in Psychology, 29*, 1–5. https://doi.org/10.1016/j.copsyc.2018.10.006

Posner, J., Song, I., Lee, S., Rodriguez, C. I., Moore, H., Marsh, R., & Blair Simpson, H. (2017). Increased functional connectivity between the default mode and salience networks in unmedicated adults with obsessive-compulsive disorder. *Human Brain Mapping, 38*(2), 678–687.

Raichle, M. (2015). The brain's default mode network. *Annual Review of Neuroscience, 38*(July), 433–447. https://doi.org/10.1146/annurev-neuro-071013-014030

Sato, T., Watanabe, K., Thompson, K., & Schall, J. (2003). Effect of target-distractor similarity on FEF visual selection in the absence of the target. *Experimental Brain Research, 151*(3), 356–363.

Schimmelpfennig, J., Topczewski, J., Zajkowski, W., & Jankowiak-Siuda, K. (2023). The role of the salience network in cognitive and affective deficits. *Frontiers in Human Neuroscience, 17*, 1133367. https://doi.org/10.3389/fnhum.2023.1133367

Smallwood, J., Bernhardt, B. C., Leech, R., Bzdok, D., Jefferies, E., & Margulies, D. S. (2021). The default mode network in cognition: a topographical perspective. *Nature Reviews Neuroscience, 22*(8), 503–513.

Sterzer, P., Voss, M., Schlagenhauf, F., & H. A. (2019). Decision-making in schizophrenia: A predictive-coding perspective. *NeuroImage, 190*, 133–143. https://doi.org/10.1016/j.neuroimage.2018.05.074

Tegelbeckers, J., Bunzeck, N., Duzel, E., Fletchtner, H., & Krauel, K. (2015). Altered salience processing in attention deficit hyperactivity disorder. *Human Brain Mapping, 36*(6), 2049–2060. https://doi.org/10.1002/hbm.22755

Vossel, S., Geng, J., & Fink, G. (2014). Dorsal and ventral attention systems: Distinct neural circuits but collaborative roles. *The Neuroscientist, 20*(2), 150–159. https://doi.org/10.1177/1073858413494269

Wang, W., Shen, J., Cheng, M., & Shao, L. (2019). AnIterative and cooperative top-down and bottom-up inference network for salient object detection. In *Proceedings of the IEEE/CVF conference on computer vision and pattern recognition* (pp. 5968–5977). CVF.

Yerys, B. E., Tunç, B., Satterthwaite, T. D., Antezana, L., Mosner, M. G., Bertollo, J. R., ... & Herrington, J. D. (2019). Functional connectivity of frontoparietal and salience/ventral attention networks have independent associations with co-occurring attentiondeficit/hyperactivity disorder symptoms in children with autism. *Biological Psychiatry: Cognitive Neuroscience and Neuroimaging, 4*(4), 343–351.

Chapter 10
Working Memory

A discussion of working memory is critical to the understanding of how attention is regulated. Working memory is a key component for understanding the role of the central executive in the regulation of attention. This is because any concept of top-down regulation of attentional resources, i.e., the imposition of a central executive who accesses past experiences and their related value determinations on the allocation of attention resources, is crucial to understanding how attention is deployed. Top-down regulation of attention initially requires that the individual has retained in memory a multiplicity of environmental stimuli and their related value determinations, and once reencountered in the environment, these memories are triggered and brought forward into working memory to be considered for action. Therefore, there is a clear connection between working memory and attention. In fact, for many situations where attention is employed, working memory is a key component of the process. While conceptually independent and subserved by different networks, a network understanding of attention would require a recognition that working memory is a necessary, and likely non-dissociable component of attention.

The earliest use of the term working memory occurred as part of the computer science literature in 1956 (Newell, 1956). In cognitive neuroscience, the term working memory was proposed by Baddeley (1992) to describe a brain network responsible for the temporary storage and manipulation of the information necessary for language comprehension, learning, and reasoning, which are complex cognitive tasks. This system was conceptualized due to the known limitations of humans on the amount of information that can be available for action at any one time. This is compared to the rather more vast amount of information that people can access when the right environmental stimuli are encountered. As in other things in the cognitive neurosciences, the meaning of the concept working memory (WM) depends on the theory in which the concept figures. The definitions differ primarily in the

T. Wasserman, L. D. Wasserman, *Disorders of Attention*, Neural Network Model: Applications and Implications, https://doi.org/10.1007/978-3-031-78732-4_10

applicable assumptions they include. Included in these are whether or not WM consists of multiple storage modules, and to what extent it includes long-term memory (Cowan, 2017). Cowan (2013) stated that working memory can be defined as the small amount of information that can be held in an especially accessible state and used in cognitive tasks. Other definitions revolve around the idea that working memory is a type of short-term memory that stores information momentarily during the completion of cognitive tasks, including comprehension, problem-solving, reasoning, and learning. Almost unique to working memory is the idea that its employment does not cause any changes in the neural network architecture of the brain since it is short-lived and momentary. Classically, working memory is described as comprised of three components. These are the phonological loop (or the verbal working memory), visuospatial sketchpad (the visual–spatial working memory), and the central executive who involves the attentional control system (Baddeley & Hitch, 1974). They describe working memory as a control system with limited information storage and processing capabilities. Baddeley (2007) provided a detailed description when he described working memory as a limited capacity system that temporarily stores and processes information for use in guiding behavior. Its three primary components include a domain-general central executive and two subsystems for the temporary storage and rehearsal of modality-specific phonological and visuospatial information. The central executive is an attentional controller responsible for oversight and coordination of the subsidiary systems. Its primary functions are focusing attention, dividing attention among concurrent tasks, and providing an interface between working memory and long-term memory. The phonological subsystem is responsible for the temporary storage and rehearsal of verbal material, whereas the visuospatial subsystem provides this function for nonverbal visual and spatial information. Baddeley indicated that extensive neuropsychological, neuroanatomical, neuroimaging, and factor analytic investigations support the distinct functioning of the two subsystems, their storage and rehearsal components, and the domain-general central executive. Finally, Baddeley further modified his definition of working memory in 1983 when he added that working memory is a temporary memory storage in connection with the performance of other cognitive tasks such as reading or problem-solving (Baddeley, 1983).

The neural network aspects of working memory have been outlined. Research indicates that working memory can be viewed as neither a unitary nor a dedicated system. Its component parts are recruited depending on a number of properties including the nature of the task at hand and the relative reward salience of that specific task. A network of brain regions, including the prefrontal cortex (PFC), is essential for the active maintenance of internal representations that are necessary for goal-directed behavior. Therefore, working memory is not localized to a single brain region but probably is an emergent property of the functional interactions between the PFC and the rest of the brain (Esposito, 2007).

Working Memory and Attention

The relationship between attention and working memory is complex and not clearly established in the literature. It is accurate to say that working memory and attention constitute two significant cognitive systems that are closely related, though they have been studied as separate constructions. Studying them separately would have, at one time, represented the state of the art and it was a standard practice to conceptualize each of these systems as standalone and believed that they could be assessed independently of their interactions with other networks, or the tasks used for assessment. We have argued, in agreement with Baddeley, that such an independent analysis of isolated elements of executive function is impossible as they all interact with, and are dependent upon, each other to produce the final result (working memory). We have also argued that the accurate assessment of working memory cannot be a globally representative measure as the assessment of working memory is not possible without considering the task used in the assessment (Wasserman & Wasserman, 2013). An example of this would occur in the situation where an individual was trying to remember a particular face. The network involved here would include the fusiform face area (FFA), a visual region that is selective for viewing faces (Kanwisher et al., 1997). The network of brain regions that was most consistently correlated with the FFA during the delay period and hence associated with the active maintenance of the represented stimulus were the prefrontal and parietal cortex, which suggest that higher-order association cortices interact with posterior sensory regions to facilitate the active maintenance of a sensory percept. Similarly, language-related visual association areas involved in the maintenance of words (as described earlier), in the absence of visual input, exhibit increased functional connectivity with the PFC but exclude the fusiform face area (Esposito, 2007). This demonstrates that the underlying components of the working memory system will change depending on the task confronting the individual.

Nevertheless, attention plays a crucial role in controlling activity, maintenance, and manipulation of working memory representations, and working memory representations play a critical role in regulating attention. There are theories that refer to attention as a resource responsible for the limited working memory capacity, while other theories refer to attention as a selection mechanism. These selection models do not, as a rule, stress working memory‘s limited capacity (Angelopoulou & Drigas, 2021). As we progress through this book, it will be evident that we, when modeling the role of working memory in attention, fall into this later group. That is, we posit two major roles for the relationship of working memory and attention. The first refers to attentional selection of the contents of working memory, controlled by mechanisms of filtering out irrelevant stimuli, and removing no-longer relevant representations from working memory. Within working memory contents, a single item is often selected as the focus of attention for processing. We shall look at how that is done in a separate chapter. The second area refers to the role of working memory in cognitive control. Working memory contributes to controlling perceptual attention because it holds templates for targets of perceptual selection, and related controlling action, by holding task sets to implement and individual's current goals (Oberauer, 2019).

Attention and Active Working Memory Capacity

As we have discussed, working memory (WM) is an active and evolving system that allows a person to maintain simultaneous active representations of information over a period of time to be used for future access or manipulation in service of specific goals or actions. Attention is thought to underpin many WM processes. One area of study is on the boundaries of WM function, specifically its capacity. Capacity limits vary between individuals and are to some extent contingent upon the properties of the representations being maintained and the current objective. It is generally accepted that WM capacity is confined to roughly three to five items (Lintz & Johnson, 2021). Obviously, some representations must exit WM at some point. People modify the operational WM set, prioritizing the selection of new items when we determine that current items may now be resolved or otherwise made irrelevant. Attention plays a role in this process.

Several mechanisms have been proposed for the role of attention in working memory. One role is that attention selects certain representations to prioritize over others and accomplishes this either by selecting representations to discard or selecting those to retain, or perhaps a combination of both. As the focus of this book is on neural networks, we would be interested in ascertaining whether these processes, discarding or retaining, are the work of one network or two. In both cases, it seems intuitive that if information is no longer relevant or desired, one strategy would be to focus elsewhere to avoid it. In the WM context, the most likely targets of one's attention would seemingly be on the non-removed, or retained, items. This interpretation is consistent with findings that the removal of some items increases the representation strength for the non-removed items and/or makes them more accessible (Lintz & Johnson, 2021). Although it looks likely that recently discarded information would be more readily available than information discarded some time ago (Lewis-Peacock et al., 2018).

Directing mental, or reflective, attention toward an item in WM is often referred to as the cognitive process of "refreshing," which has been found to strengthen a refreshed item's representation relative to the other items in that WM set. This heightened attention to the representation increases its availability and leads to improved long-term memory relative to items that were not refreshed (Raye et al., 2007).

While there are arguments that refreshing and removal are distinct processes subserved by slightly differing networks, the net effect on working memory might be better understood and an integrated process. After all, neither process occurs in isolation from the other. Specifically, perhaps it would be preferable simply to reframe both refreshing and removal as complementary aspects of an overarching process of reallocation of a finite resource in the form of mental or reflective attention (Chun & Johnson, 2011). When a subset of the items currently held in WM are identified as more relevant than others, attention is shifted toward the relevant item(s) in an effort to strengthen their representations and prolong successful maintenance. As a result, attention is likewise withdrawn from other items. In this conceptualization, refreshing and removal still occur as *tasks* people perform or strategies used to perform them, but the primary core process underlying them both is the same.

Attention, Working Memory, and Visual Search

A good deal of the research that suggests a role for working memory in attention processing comes from the research on visual search of the environment for both primates and, more recently, humans. Several influential theories of attention posit that attention is controlled during visual search by a form a pattern recognition that is brought into and maintained in visual working memory during search (Woodman et al., 2007). In the extant research, this pattern recognition is frequently referred to as a target template. The nature of the template is not necessarily a pictorial representation but is nevertheless linked with visual identification. The idea is rather straightforward. A form of representation of an environmental object that is encountered for the first time is learned, stored in long-term memory, and when encountered again that "template" is brought forward into working memory, to be worked upon or ignored, whatever the situation requires. In actuality, this is not entirely correct for complex environmental stimuli, but by the time the information reaches, working memory is likely, at least for the visual component, to be in template form. There is research that outlines the network components of this process. For example, neurons in the prefrontal cortex retain target representations, and this sustained activity feeds back to neurons in posterior visual areas that perform primary perceptual analysis the target features (Goldman-Rakic, 1996). This model notes that the dorsal and ventral streams of the posterior visual system converge on superior and inferior parts of the lateral prefrontal cortex. The inferior portion of the lateral prefrontal cortex maintains working memory representations of objects. Because these object representations are essentially separated from any specific spatial coordinates, they permit flexible behavioral interactions with the objects in the environments that vary widely across space. More recent models of memory would suggest that these visual representations are stored elsewhere and brought to working memory when reencountered. The important take away from this line of research is that working memory is essential in directing attention. We would argue that this is because the "targets" represent potential value and must be analyzed to determine their specific value in the situation currently encountered.

This entire process does not seem to be the work of one dedicated network but of at least two. Research indicates that the target template is brought to visual working memory when the target changes frequently. When a search target remains constant, it is not held in working memory (Woodman et al., 2007). By focusing on the activity of the visual activity in this network, studies have found that the efficiency of attentional deployment during search is significantly influenced by the recency of a change in target identity. In a variable-target condition, executive control over search performance is implemented by maintaining, in a top-down manner, a target template in visual working memory. Maintaining such a representation helps to suppress attentional deployment to targets from previous trials, or competing nonessential targets in the same space that are now distractors, and to enable accurate performance of the search task in the face of concurrently maintenance of the other information in working memory. Hypothetically, inefficiency in generating these templates or holding them in memory would result in the person appearing to not attend. In this case, the difficulty might be related to slow template generation or failure to hold the template in memory.

Working Memory Deficits in Attention Deficit Hyperactivity Disorder

Given the fact that working memory is a key factor of at least executively managed (top-down) allocation of attentional resources, it would stand to reason that problems with working memory would be found in populations of individuals with attention deficit hyperactivity disorder (ADHD), which they are. In fact, they are somewhat ubiquitous (Alderson et al., 2013; Kapser et al., 2012; Martinussen et al., 2005). Converging evidence indicates that children with ADHD are impaired in all three components of working memory, with the largest deficits found in the domain-general central executive (CE) system, followed by visuospatial (VS) storage/rehearsal and then phonological (PH) storage/rehearsal subsystems) (Kofler et al., 2009). The central executive component of working memory is also functionally related to the excess motor activity (i.e., hyperactivity), which is a hallmark and key diagnostic feature of ADHD (Rapport et al., 2009).

There are in fact models that suggest a direct causal relationship between working memory deficits and ADHD (Kofler et al., 2009). Their model posits two conditions that account for the attentive behavior deficits in children with ADHD. These are (1) placing demands on central executive processing, the effect of which is evident under even low cognitive loads, and (2) exceeding storage/rehearsal capacity, which has similar effects on children with ADHD and typically developing children but occurs at lower cognitive loads for children with ADHD. Just how this process occurs is not clear and the amount needed to overload a system varies from individual to individual. Arguably, those individuals with storage capacity deficits would not be inattentive per sea, but, rather, they would be inefficient in keeping all relevant factors of certain situations in contemporaneous memory. They would look inattentive because they would miss certain cues but typical treatments designed to focus attention might be ineffective.

In comparison with the direct causality models, both auditory working memory issues and visual working memory issues have been found to be associated with attention deficit hyperactivity disorder (Raiker et al., 2019). The literature clearly establishes that ADHD and working memory are associated and that one does not cause the other. There is also considerable research that suggests that the differing types of ADHD recognized diagnostically are associated with differing types of working memory disruptions. For example, it has been hypothesized that the ADHD-inattentive subtype may constitute a different type of attention disorder compared to the ADHD-hyperactive type and ADHD-combined type subtypes (Schweitzer et al., 2006). In this model, both the hyperactive and combined types are conceptualized as, centrally, exhibiting deficits in behavioral inhibition, which in turn leads to impairments in executive functioning and working memory (WM). The inattentive type ADHD is a fundamentally distinct disorder with problems in attention arising from non-inhibitory mechanisms. This model has received mixed support in the literature.

While it might be accurate to say that those individuals who present themselves in clinical settings frequently have ADHD and learning challenges. There might be individuals who meet the behavioral definition of ADHD but do not manifest working memory challenges (Liebel & Nelson, 2017).

We will say at the outset that in most of the studies that were reviewed, children with any form of ADHD diagnosis were lumped together for statistical analysis. This obscures the possibility that there is a subset of children who are diagnosed with ADHD and do not have working memory issues. There is some research to support this hypothesis. This research suggests that there is a fundamental heterogeneity among ADHD patients that yields different findings depending upon the proportion of patients with impaired or intact working memory represented in any given sample (Mattfield et al., 2016). Research has also suggested that half of ADHD patients are unimpaired on any particular measure of executive function (Biederman, 2004). This research suggests differing but overlapping pathways for attention and working memory and is about half the cases reviewed that there was no damage to the working memory network. This finding highlights the concern that such heterogeneity among ADHD patients can lead to misleading and obscuring findings.

The fact that there are differing networks for working memory and attention is entirely consistent with network modeling. Both systems arise in relation to differing task demands placed upon the individual. One system, attention, is an initially bottom-up system that is inborn and responds to novelty. The other system develops in relation to the individual's survival requirement to respond to flight or fight targets. From birth, these systems become dependent upon each other to produce significant aspects of childhood and adult attention allocation. Complex human behavior requires integrated networks. It is absolutely clear that both networks (attention and working memory) must work together in an integrated fashion. Problems can arise when either network is inefficient, but research suggests that individuals with deficits in both networks experience additive deficits that impact functioning to a greater extent (Maehler & Schuchardt, 2016).

Notwithstanding the above, the extant research does point to working memory deficits in the ADHD population and leaves open the question as to which of the areas of weakness (attention or executive management) is causing the behavior leading to the diagnosis.

References

Abramovitch, A., Dar, R., Hermesh, H., & Schweiger, A. (2011). Comparative neuropsychology of adult obsessive-compulsive disorder and attention deficit/hyperactivity disorder: Implications for a novel executive overload model of OCD. *Journal of Neuropsychology*. https://doi.org/10.1111/j.1748-6653.2011.02021

Alderson, M., Kasper, L., Hudec, K., & Patros, C. (2013). Attention-deficit/hyperactivity disorder (ADHD) and working memory in adults: A meta-analytic review. *Neuropsychology, 27*(3), 287–302. https://doi.org/10.1037/a0032371

Alderson, T., Bokde, A., Kelso, J., Mcquire, L., & Coyle, D. (2020). Metastable neural dynamics underlies cognitive performance across multiple behavioural paradigms. *Human Brain Mapping, 41*(12), 3212–3234. https://doi.org/10.1002/hbm.25009

Allan, F. M., & Anthony, B. J. (1991). Analysis of the elements of attention: A neuropsychological approach. *Neuropsychology Review, 2*, 109–145.

American Psychiatric Association. (1968). *Diagnostic and statistical manual of mental disorders (2nd edition; DSM-II)*. American Psychiatric Association.

American Psychiatric Association. (2013). *The diagnostic and statistical manual of mental disorders (5th edition; DSM-5)*. American Psychiatric Association.

Andersen, B., Lauremt, P., & Yantis, S. (2011). Value-driven attentional capture. *PNAS Biological Science, 108*(25), 10367–10371. https://doi.org/10.1073/pnas.1104047108

Anderson, B. (2011). There is no such thing as attention. *Frontiers in Psychology, 2*, 246. https://doi.org/10.3389/fpsyg.2011.00246

Anderson, B. (2017). Reward processing in the value-driven attention network: Reward signals tracking cue identity and location. *Social Cognitive and Affective Neuroscience, 12*(3), 461–467. https://doi.org/10.1093/scan/nsw141

Anderson, P., He, X., Buehler, C., Teney, D., Johnson, M., & Gould, S. (2018). Bottom-up and top-down attention for image captioning and visual question answering. In *CVPR proceedings of the IEEE conference on computer vision and pattern recognition* (pp. 6077–6086).

Andoh, J., Matsushita, R., & Zatorre, R. J. (2015). Asymmetric interhemispheric transfer in the auditory network: Evidence from TMS, resting-state fMRI, and diffusion imaging. *Journal of Neuroscience, 35*(43), 14602–14611.

Andrade, M. N. (2023). *History of attention*. Retrieved from LibreTexts Social Sciences: https://socialsci.libretexts.org/Bookshelves/Psychology/Cognitive_Psychology/Cognitive_Psychology

Angelopoulou, E., & Drigas, A. (2021). Working memory, attention and their relationship: A theoretical overview. *Research, Society and Development, 10*(5), e46410515288.

Armel, K., Beaumel, A., & Rangel, A. (2006). Biasing simple choices by manipulating relative visual attention. *Judgment and Decision Making, 3*(5), 396–403.

Arnett, A., Rhodes, C., & Rutter, T. (2021). Reduced error recognition explains post-error slowing differences among children with attention deficit hyperactivity disorder. *Journal of the International Neuropsychological Society, 28*(8), 810. https://doi.org/10.1017/S1355617721001065

Arsten, A., & Rubia, K. (2012). Neurobiological circuits regulating attention, cognitive control, motivation, and emotion: Disruptions in neurodevelopmental psychiatric disorders. *Journal of the American Academy of Child and Adolescent Psychiatry, 51*(4), 356–367. https://doi.org/10.1016/j.jaac.2012.01.008

Awh, E., Belopolsky, A., & Theeuwes, J. (2012). Top-down versus bottom-up attentional control: A failed theoretical dichotomy. *Trends in Cognitive Sciences., 16*(8), 437–443. https://doi.org/10.1016/j.tics.2012.06.010

Baddeley, A. (1992). Working memory. *Science, 255*(5044), 556–559.

Baddeley, A. D. (1983). Working memory. Philosophical Transactions of the Royal Society of London. *B, Biological Sciences, 302*(1110), 311–324.

Baddeley, A. D. (2007). Working memory, thought, and action (Vol. 8). Oxford University Press.

Baddeley, A., & Hitch, G. (1974). Working memory. In I. G. Bower (Ed.), *Recent advances in learning and motivation, 8* (pp. 47–49). Academic. https://doi.org/10.1016/S0079-7421(08)60452-1

Balkenius, C., Förster, A., Johansson, B., & Thorsteinsdottir, V. (2008). Anticipation in attention. In I. G. Pezzulo, M. Butz, C. Castelfranchi, & R. Falcone (Eds.), *The challenge of anticipation* (Lecture notes in computer science) (Vol. 5225). Springer. https://doi.org/10.1007/978-3-540-87702-8_4

Ball, G., et al. (2014). Rich-club organization of the newborn human brain. *Proceedings of the National Academy of Sciences of the United States of America, 111*, 7456–7461. https://doi.org/10.1073/pnas.1324118111

Barker, H., Munro, J., Orlov, N., Morgenroth, E., Moser, J., Eysenck, M., & Allen, P. (2018). Worry is associated with inefficient functional activity and connectivity in prefrontal and cingulate cortices during emotional interference. *Brain and Behavior, 8*(12), e01137. https://doi.org/10.1002/brb3.1137

Barkley, R., & Fischer, M. (2010). The unique contribution of emotional impulsiveness to impairment in major life activities in hyperactive children as adult. *Journal of the American Academy of Child and Adolescent Psychiatry, 49*(5), 503–513.

Barnes, R., & Jones, M. (2000). Expectancy, attention, and time. *Cognitive Psychology., 41*(3), 254–311. https://doi.org/10.1006/cogp.2000.0738

Beck, D., & Kastner, S. (2005). Stimulus context modulates competition in human extrastriate cortex. *Nature Neuroscience, 8*(8), 1110–1116. https://doi.org/10.1038/nn1501

Beck, D., & Krastner, S. (2009). Top-down and bottom-up mechanisms in biasing competition in the human brain. *Vision Research, 49*(10), 1154–1165. https://doi.org/10.1016/j.visres.2008.07.012

Beckmann, C. F. (2005). Investigations into resting-state connectivity using independent component analysis. *Philosophical Transactions of the Royal Society B: Biological Sciences, 360*(1457), 1001–1013.

Benes, F. (2000). Emerging principles of altered neural circuitry in schizophrenia. *Brain Research Review, 31*(2–3), 251–289.

Berger, A., et al. (2021). How is the norepinephrine system involved in the antiepileptic effects of Vagus nerve stimulation? *Frontiers in Neuroscience, 15*. https://doi.org/10.3389/fnins.2021.790943

Berggren, N., & Derakshan, N. (2013). Attentional control deficits in trait anxiety: Why you see them and why you don't. *Biological Psychology, 92*(3), 440–446. https://doi.org/10.1016/j.biopsycho.2012.03.007

Beucke, J., Sepulcre, J. E., Sebold, M., Kathmann, N., & Kaufmann, C. (2014). Default mode network subsystem alterations in obsessive-compulsive disorder. *British Journal of Psychiatry, 205*, 376–382. https://doi.org/10.1192/bjp.bp.113.137380

Biederman, J. C. (2004). Impact of executive function deficits and attention-deficit/hyperactivity disorder (ADHD) on academic outcomes in children. *Journal Of Clinical and Consulting Psychology, 72*(5), 757–776.

Biederman, J., Monuteaux, M., Doyle, A., Seidman, L., Wilens, T., Ferrero, F., et al. (2004). Impact of executive function deficits and attention-deficit/hyperactivity disorder (ADHD) on academic outcomes in children. *Journal of Consulting and Clinical Psychology, 72*(5), 757–766. https://doi.org/10.1037/0022-006X.72.5.757

Blum, K., Li-Chuan Chen, A., Braverman, E., Comings, D., Chen, T., & Arcuri, V. (2008). Attention-deficit-hyperactivity disorder and reward deficiency syndrome. *Neuropsychiatric Disease and Treatment, 4*, 5. https://doi.org/10.2147/ndt.s2627

Bogacz, R., Hu, P., & Cohen, J. (2010). Do humans produce the speed–accuracy trade-off that maximizes reward rate? *Quarterly Journal of Experimental Psychology, 63*, 5. https://doi.org/10.1080/17470210903091643

Bora, E. (2020). Meta-analysis of neurocognitive deficits in unaffected relatives of obsessive-compulsive disorder (OCD): Comparison with healthy controls and patients with OCD. *Psychological Medicine, 50*(8), 1257–1266. https://doi.org/10.1017/S0033291720001634

Borji, A. (2012). Boosting bottom-up and top-down visual features for saliency estimation. In *IEEE conference on computer vision and pattern recognition* (pp. 438–445). IEEE.

Bossong, M., & Kahn, R. (2016). The salience of reward. *JAMA Psychiatry, 73*(8), 777–778. https://doi.org/10.1001/jamapsychiatry.2016.1134

Braver, T. (2012). The variable nature of cognitive control: A dual mechanisms framework. *Trends in Cognitive Sciences, 16*(2), 106–113. https://doi.org/10.1016/j.tics.2011.12.010

Broadbent, D. (1958). *Perception and communication*. Oxford University Press.

Bronco, P. (2020). Mapping language with resting-state functional magnetic resonance imaging: A study on the functional profile of the language network. *Huma Brain Mapping, 41*(2), 545–560. https://doi.org/10.1002/hbm.24821

Bunford, N., Hamori, G., Nemoda, Z., Angyal, N., Fiath, R. U., & Paszthy, B. (2023). The domain-variant indirect association between electrophysiological response to reward and ADHD presentations is moderated by dopaminergic polymorphisms. *Comprehensive Psychiatry, 124*, 152389. https://doi.org/10.1016/j.comppsych.2023.152389

Bush, G. (2010). Attention-deficit/hyperactivity disorder and attention networks. *Neuropsychopharmacology Reviews, 35*, 278–300. https://doi.org/10.1038/npp.2009.120

Cabral, L., Zubiaurre, L., Wild, C., Linke, A., & Cusack, R. (2019). Category-selective visual regions have distinctive signatures of connectivity in neonates. *BioRxiv*, 675421. https://doi.org/10.1101/675421

Carrasco, M. (2011). Visual attention: The past 25 years. *Vision Research, 51*(13), 1484–1525.

Chun, M., & Johnson, M. (2011). Memory: Enduring traces of perceptual and reflective attention. *Neuron, 72*, 520–535. https://doi.org/10.1016/j.neuron.2011.10.026

Churchland, M. M., & Shenoy, K. V. (2007). Temporal complexity and heterogeneity of single-neuron activity in premotor and motor cortex. *Journal of Neurophysiology, 97*, 4235–4257.

Cieslik, E., Mueller, V., Eickhoff, C., Langner, R., & Eickhoff, S. (2015). Three key regions for supervisory attentional control: Evidence from neuroimaging meta-analyses. *Neuroscience & Biobehavioral Reviews, 48*, 22–34.

Clark, L. C.-F. (2017). Three approaches to understanding and classifying mental disorder: ICD-11, DSM-5, and the National Institute of Mental Health's research domain criteria (RDoC). *Sage Journals Psychological Science in the Public Interest, 18*, 72. https://doi.org/10.1177/1529100617727266

Coghill, D., & Banaschewski, T. (2009). The genetics of attention-deficit/hyperactivity disorder. *Expert Review of Neurotherapeutics, 9*(10), 1547–1565. https://doi.org/10.1586/ern.09.78

Cohen, R. A., Malloy, P. J. M., & Paul, R. (2014). Disorders of attention. In M. Parsons & T. Hammeke (Eds.), *Clinical neuropsychology: A pocket handbook for assessment* (pp. 463–497). American Psychological Association. https://doi.org/10.1037/14339-021

Corbetta, M., & Shulman, G. (2002). Control of goal-directed and stimulus-driven attention in the brain. *Nature Reviews Neuroscience, 3*, 201–215.

Corlett, P., Honey, G., & Fletcher, P. (2016). Prediction error, ketamine and psychosis: An updated model. *Journal of Psychopharmacology, 30*, 1145–1155. https://doi.org/10.1177/0269881116650087

Costa-Dias, T., Wilson, V., Bathula, D., Iyer, S., Mills, K., Thurlow, B., & Stevens, C. (2013). Reward circuit connectivity relates to delay discounting in children with attention-deficit/hyperactivity disorder. *European Neuropsychopharmacology, 23*(1), 33–45. https://doi.org/10.1016/j.euroneuro.2012.10.015

Coull, J., Sahakian, B., & Hodges, J. (1996). The alpha (2) antagonist idazoxan remediates certain attentional and executive dysfunction in patients with dementia of frontal type. *Psychopharmacology (Berlin), 123*(3), 239–249.

Cowan, N. (2013). Short-term and working memory in childhood. The Wiley handbook on the development of children's memory, *1*, 202–229.

Cowan, N. (2017). The many faces of working memory and short-term storage. *Psychonomic Bulletin & Review, 24*, 1158–1170.

Cuthburt, B. (2022). Research domain criteria (RDoC): Progress and potential. *Current Directions of Psychological Science, 31*(2), 107–114. https://doi.org/10.1177/09637214211051363

de la Pena, M., Pan, M., Thai, C., & Alisso, T. (2020). Attention-deficit/hyperactivity disorder predominantly inattentive subtype/presentation: Research Progress and translational studies. *Brain Science, 10*(5), 292. https://doi.org/10.3390/brainsci10050292

Derryberry, D., & Reed, M. (2002). Anxiety-related attentional biases and their regulation by attentional control. *Journal of Abnormal Psychology, 111*(2), 225–236. https://doi.org/10.1037/0021-843X.111.2.225

Deserno, M., Bathelt, J., & Groenman, A. E. (2023). Probing the overarching continuum theory: Data-driven phenotypic clustering of children with ASD or ADHD. *European Child Adolesc Psychiatry, 32*, 1909–1923. https://doi.org/10.1007/s00787-022-01986-9

Dickstein, S., Bannon, K., Castellanos, F., & Milha, M. (2006). The neural correlates of attention deficit hyperactivity disorder: An ALE meta-analysis. *Journal of Child Psychology and Psychiatry, 47*, 1051–1062. https://doi.org/10.1111/j.1469-7610.2006.01671.x

Dong, G., Lin, X., & Potenza, N. (2015). Decreased functional connectivity in an executive control network is related to impaired executive function in internet gaming disorder. *Progress in Neuro-Psychopharmacology and Biological Psychiatry, 57*, 76–85. https://doi.org/10.1016/j.pnpbp.2014.10.012

Dosenbach, N., Fair, D., Cohen, A. L., Schlaggar, B., & Petersen, S. (2008). A dual-networks architecture of top-down control. *Trends in Cognitive Sciences, 12*(3), 99–105. https://doi.org/10.1016/j.tics.2008.01.001

Drigas, A., & Karyotaki, M. (2019). Attention and its role: Theories and models. *International Journal of Emerging Technologies in Learning., 14*(12), 169–182.

Duménieu, M., Marquèze-Pouey, B., Russier, M., & Debanne, D. (2021). Mechanisms of plasticity in subcortical visual areas. *Cells, 10*(11), 3162. https://doi.org/10.3390/cells10113162

Durston, S., van Belle, J., & de Zeeuw, P. (2011). Differentiating Frontostriatal and Frontocerebellar circuits in attention-deficit/hyperactivity disorder. *Biological Psychiatry, 69*(12), 1178–1184. https://doi.org/10.1016/j.biopsych.2010.07.037

Ebrahimpour, M., et al. (2019). Ventral-dorsal neural networks: Object detection via selective attention. In *2019 IEEE Winter Conference on Applications of Computer Vision (WACV)* (pp. 986–994). IEEE. https://doi.org/10.1109/WACV.2019.00110

Englemann, J., Damaraju, E., Padmala, S., & Pessoa, L. (2009). Combined effects of attention and motivation on visual task performance: Transient and sustained motivational effects. *Fronteirs in Human Neuroscience, 3*, 4. https://doi.org/10.3389/neuro.09.004.2009

Esposito, M. (2007). From cognitive to neural models of working memory. *Philosophical Transactions of the Royal Society of London, 362*(1481), 761–772. https://doi.org/10.1098/rstb.2007.2086

Esposito, R., Cieri, F., Chiacchiaretta, P., Cera, N., Lauriola, M., Di Giannantoni, M., et al. (2018). Modifications in resting state functional anticorrelation between default mode network and dorsal attention network: Comparison among young adults, healthy elders and mild cognitive impairment patients. *Brain Imaging and Behavior, 12*(1), 127–141. https://doi.org/10.1007/S11682-017-9686-Y

Estes, B., & Polnick, B. (2012). Examining motivation theory in higher education: An expectancy theory analysis of tenured faculty productivity. *International Journal of Management, Business, and Administration., 15*(1), 1–7.

Evarts, E. V. (1868). Relation of pyramidal tract activity to force exerted during voluntary movement. *Journal of Neurophysiology, 31*, 14–27.

Everling, S., Tinsley, C., Gaffan, D., & Duncan, J. (2006). Selective representation of task-relevant objects and locations in the monkey prefrontal cortex. *European Journal of Neuroscience, 23*, 2197–2214. https://doi.org/10.1111/j.1460-9568.2006.04736.x

Eysenck, M. (1992). *Anxiety: The cognitive perspective*. Erlbaum.

Eysenck, M. W., & Calvo, M. (1992). Anxiety and performance: The processing efficiency theory. *Cognition and Emotion, 6*, 409–434.

Eysenck, M., Moser, J., Derakshan, N., & Hepsomali, P. A. (2022). A neurocognitive account of attentional control theory: How does trait anxiety affect the brain's attentional networks? *Cognition and Emotion, 37*(2), 220–237. https://doi.org/10.1080/02699931.2022.2159936

Failing, M., & Theeuwes, J. (2018). Selection history: How reward modulates selectivity of visual attention. *Psychonomic Bulletin & Review, 25*, 514–538. https://doi.org/10.3758/s13423-017-1380-y

Fan, F., Liao, L., Lei, T., Zhao, T., Xia, M., Men, W., et al. (2021). Development of the default-mode network during childhood and adolescence: A longitudinal resting-state fMRI study. *NeuroImage, 226*, 1. https://doi.org/10.1016/j.neuroimage.2020.117581

Fitzgerald, K., Stern, E., Angstadt, M., Nicholson-Muth, K., Maynor, M., Welsh, R. C., et al. (2010). Altered function and connectivity of the medial frontal cortex in pediatric obsessive-compulsive disorder. *Biological Psychiatry, 68*, 1039–1047. https://doi.org/10.1016/j.biopsych.2010.08.018

Fortunato, C., Bennasar-Vázquez, J., Park, J., Chang, J., Miller, L., Dudman, J., et al. (2024). Nonlinear manifolds underlie neural population activity during behaviour. *bioRxiv [Preprint]*. https://doi.org/10.1101/2023.07.18.549575

Friston, K. (2008). Hierarchical models in the brain. *PLoS Computational Biology, 4*, e1000211.

Friston, K., Adams, R., Perrinet, L., & Breakspear, M. (2012). Perceptions as hypotheses: Saccades as experiments. *Frontiers in Psychology, 3*, 151.

Fuster, J. (1997). Network memory. *Trends in Neuroscience, 20*(10), 451–459. https://doi.org/10.1016/S0166-2236(97)01128-4

Gallego, J., Perich, M., Naufel, S., Ethier, C., Solla, S., & Miller, E. (2018). Cortical population activity within a preserved neural manifold underlies multiple motor behaviors. *Nature Communications, 9*, 4233. https://doi.org/10.1038/s41467-018-06560-z

Gerstner, W., Kistler, W., Naud, R., & Paninsk, L. (2014). Neuronal Populations. In W. Gerstner, W. Kistler, R. Naud, & L. Paninsk (Eds.), *Neuronal dynamics from single neurons to networks and models of cognition* (pp. 291–322). Cambridge University Press.

Girault, J., Veatch, O., & Won, H. (2024). Etiologic heterogeneity, pleiotropy, and polygenicity in behaviorally defined intellectual and developmental disabilities. *Journal of Neurodevelopmental Disorders, 16*, 8. https://doi.org/10.1186/s11689-024-09526-z

Goldman-Rakic, P. (1996). Regional and cellular fractionation of working memory. *Proceedings of the National Academy of Sciences of the United States of America, 93*(24), 13473–13480.

Grizenko, N., Paci, M., & Joober, R. (2010). Is the inattentive subtype of ADHD different from the combined/hyperactive subtype? *Journal of Attention Disorders, 13*(6), 649–657. https://doi.org/10.1177/1087054709347200

Harvey, P. (2019). Domains of cognition and their assessment. *Dialogues in Clinical Neuroscience*, 227–237. https://doi.org/10.31887/DCNS.2019.21.3/pharvey

Hermans, E., Henckens, M., & Joëls, M. G. (2014). Dynamic adaptation of large-scale brain networks in response to acute stressors. *Trends in Neuroscience, 37*, 304–314. https://doi.org/10.1016/j.tins.2014.03.006

Hillyard, S., Vogel, E., & Luck, S. (1998). Sensory gain control (amplification) as a mechanism of selective attention: Electrophysiological and neuroimaging evidence. *Philosophical Transactions of the Royal Society of London on Biological Science, 353*(1373), 1257–1270. https://doi.org/10.1098/rstb.1998.0281

Hofman, M. (2014). Evolution of the human brain: When bigger is better. *Frontiers in Neuroanatomy, 8*, 26. https://doi.org/10.3389/fnana.2014.00015

Holland, N., Robbins, T., & Rowe, J. (2021). The role of noradrenaline in cognition and cognitive disorders. *Brain, 144*, 2243–2256. https://doi.org/10.1093/brain/awab111

Hovel, H., Viol, A., Loske, P., Merfort, L., & Vuksanović, V. (2020). Synchronization in functional networks of the human brain. *Journal of Non Linear Science, 30*, 2259–2282. https://doi.org/10.1007/s00332-018-9505-7

Humphrey, D. R., & Thompson, W. D. (1970). Predicting measures of motor performance from multiple cortical spike trains. *Science, 170*(7), 758–762.

Hyman, A. (2005). Addiction: A disease of learning and memory. *The American Journal of Psychiatry, 162*, 1414. https://doi.org/10.1176/appi.ajp.162.8.1414

Itti, L., & Baldi, P. (2006). Bayesian surprise attracts human attention. *Advances in Neural Information Processing Systems, 18*, 547.

James, H. (1890). *The principles of psychology*. Holt.

Jankowiak-Siuda, K., & Zajkowski, W. (2013). A neural model of mechanisms of empathy deficits in narcissism. *Medical Science Monitor, 19*, 934–941. https://doi.org/10.12659/MSM.889593

Jensen, C., & Steinhausen, H. (2015). Comorbid mental disorders in children and adolescents with attention-deficit/hyperactivity disorder in a large nationwide study. *ADHD Attention Deficit and Hyperactivity Disorders, 7*(1), 27–38.

Joel, D. W. (2000). The connections of the dopaminergic system with the striatum in rats and primates: An analysis with respect to the functional and compartmental organization of the striatum. *Neuroscience, 96*, 451–474.

Johnson, M., Gilga, T., & Charman, T. (2014). Annual Research Review: Infant development, autism, and ADHD – Early pathways to emerging disorders. *The Journal of Child Psychology and Psychiatry*. https://doi.org/10.1111/jcpp.12328

Kanwisher, N., McDermott, J., & Chun, M. (1997). The fusiform face area: A module in human extrastriate cortex specialized for face perception. *Journal of Neuroscience, 17*, 4302–4311.

Kapser, L., Aldrson, R., & Hudec, K. (2012). Moderators of working memory deficits in children with attention-deficit/hyperactivity disorder (ADHD): A meta-analytic review. *Clinical Psychology Review, 32*(7), 605–617.

Katsuki, M., & Constantinidis, C. (2013). Bottom-up and top-down attention: Different processes and overlapping neural systems. *The Neuroscientist, 20*(5), 509–522. https://doi.org/10.1177/1073858413514136

Katsuki, F., & Constantinidis, C. (2014). Bottom-up and top-down attention: Different processes and overlapping neural systems. *The Neuroscientist*, 509–521. https://doi.org/10.1177/1073858413514136

Katsuki, F., & Constantinidis, C. (2016). Bottom-up and top-down attention: Different processes and overlapping neural systems. *The Neuroscientist, 20*, 1–13. https://doi.org/10.1177/1073858413514136

Keller, A., et al. (2019). Paying attention to attention in depression. *Translational Psychiatry, 9*, 279. https://doi.org/10.1038/s41398-019-0616-1

Kessler, R. C., et al. (2006). The prevalence and correlates of adult ADHD in the United States: Results from the National Comorbidity Survey Replication. *Amer J Psychiatry, 163*(4), 716–723.

Kim, N., & Kastner, S. (2019). A biased competition theory for the developmental cognitive neuroscience of visuo-spatial attention. *Current Opinion in Psychology*, 29. https://doi.org/10.1016/j.copsyc.2019.03.017

Kimura, A., Yonetani, R., & Hirayami, T. (2013). Computational models of human visual attention and their implementations: A survey. *IEICE Transactions on Information and Systems, E96-D*(3), 562–578.

Koch, C., & Ullman, S. (1985). Shifts in selective visual attention: Towards the underlying neural circuitry. *Human Neurobiology, 4*(2), 19–27.

Kofler, M., Rapport, M., Boldin, J., Sarver, D., & Raiker, J. (2009). ADHD and working memory: The impact of central executive deficits and exceeding storage/rehearsal capacity on observed inattentive behavior. *Journal of Abnormal Child Psychology, 38*, 149–161. https://doi.org/10.1007/s10802-009-9357-6

Kolb, B. M. (2012). Experience and the development of the prefrontal cortex. *Proceedings of the National Academy of Science, 109*, 17186–17193.

Koziol, L. F., & Budding, D. E. (2009). *Subcortical structures and cognition: Implications for neuropsychological assessment*. Springer.

Koziol, L., & Lutz, J. (2013). From movement to thought: The development of executive function. *Applied Neuropsychology: Child, 2*(2), 104–115. https://doi.org/10.1080/21622965.2013.748386

Koziol, A., Joyce, A., & Wurglitz, G. (2014). The neuropsychology of attention: Revisiting the "Mirsky Model". *Applied Neuropsychology: Child, 3*(4). https://doi.org/10.1080/21622965.2013.870016

Kunsti, J., et al. (2014). The separation of ADHD inattention and hyperactivity-impulsivity symptoms: Pathways from genetic effects to cognitive impairments and symptoms. *Journal of Abnormal Child Psychology., 42*, 127–136. https://doi.org/10.1007/s10802-013-9771-7

Le-Cunff, A. (2022). *Is there a limit to the human brain's capacity?* Retrieved from Ness labs: https://nesslabs.com/brain-limitations

Lee, D., Quattrocki, E. K., Lee, S., Pae, C., & Yoo, S. (2021). Differential structure-function network coupling in the inattentive and combined types of attention deficit hyperactivity disorder. *PLoS One, 16*, e0260295. https://doi.org/10.1371/journal.pone.0260295

Legrain, V., Plaghki, L., & Mouraux, A. (2011). The pain matrix reloaded. *Progress in Neurobiology, 93*, 111–124. https://doi.org/10.1016/j.pneurobio.2010.10.005

Lewis-Peacock, J., Kessler, Y., & Oberauer, K. (2018). The removal of information from working memory. *Annals of the New York Academy of Sciences, 1424*(1), 33–44. https://doi.org/10.1111/nyas.13714

Liebel, S., & Nelson, J. (2017). Auditory and visual working memory functioning in college students with attention-deficit/hyperactivity disorder and/or learning disabilities. *Archives of Clinical Neuropsychology, 32*(8), 980–991. https://doi.org/10.1093/arclin/acx014

Limongi, R., et al. (2020). Glutamate and dysconnection in the salience network: neurochemical, effective connectivity and computational evidence in schizophrenia. *Biological Psychiatry, 88*, 273–281. https://doi.org/10.1016/j.biopsych.2020.01.021

Lintz, E., & Johnson, M. (2021). Refreshing and removing items in working memory: Different approaches to equivalent processes? *Cognition, 211*, 104655. https://doi.org/10.1016/j.cognition.2021

Liu, J. C., Liu, K., & Zhao, J. (2018). *Event detection via gated multilingual attention mechanism. Thirty-second AAAI conference on artificial intelligence*. Association for the Advancement of Artificial Intelligence.

Lubke, G. H., Derkes, E., van Bijsterveldt, T., & Boomsma, D. (2009). Maternal ratings of attention problems in ADHD: Evidence for the existence of a continuum. *Journal of the American Academy of Child & Adolescent Psychiatry., 48*(11), 1085–1093. https://doi.org/10.1097/CHI.0b013e3181ba3dbb

Lux, V. (2018). Epigenetic programming effects of early life stress: A dual-activation hypothesis. *Current Genomics, 19*(8), 638–652. https://doi.org/10.2174/1389202919666180307151358

Mackintosh, N. J. (2016). A theory of attention: Variations in the associability of stimuli with reinforcement. *Psychological Review, 82*(4), 276–298. https://doi.org/10.1037/h0076778

MacLeod, C., & Clarke, P. (2015). The attentional bias modification approach to anxiety intervention. *Clinical Psychological Science, 3*(1), 58–78. https://doi.org/10.1177/2167702614560749

MacLeod, C., Rutherford, E., Campbell, L., Ebsworthy, G., & Holker, L. (2002). Selective attention and emotional vulnerability: Assessing the causal basis of their association through the experimental manipulation of attentional bias. *Journal of Abnormal Psychology, 11*, 107–123.

Maehler, C., & Schuchardt, K. (2016). Working memory in children with specific learning disorders and/or attention deficits. *Learning and Individual Differences, 49*, 341–347. https://doi.org/10.1016/j.lindif.2016.05.007

Magnus, W. N. (2023). *Attention deficit hyperactivity disorder*. Retrieved from National Library of Medicine National Center for Biotechnology Information: https://www.ncbi.nlm.nih.gov/books/NBK441838/

Makovac, E., Fagioli, S., Rae, C. L., Critchley, H., & Ottaviani, C. (2020). Can't get it off my brain: Meta-analysis of neuroimaging studies on perseverative cognition. *Psychiatry Research: Neuroimaging, 295*, 111020. https://doi.org/10.1016/j.pscychresns.2019.111020

Manoliu, A., et al. (2014). Insular dysfunction within the salience network is associated with severity of symptoms and aberrant inter-network connectivity in major depressive disorder. *Frontiers in Human Neuroscience, 7*, 21. https://doi.org/10.3389/fnhum.2013.00930

Mariotti, P., Mariotti, P., Lucibello, S., De Clemente, V., Veltri, S., Janiri, D., et al. (2021). ADHD syndrome and Tic disorders: Two sides of the same coin. *The Lancet*. https://doi.org/10.2139/ssrn/3923340

Markett, S., Nothdurfter, D., Fosca, A., & Reuter, M. (2022). Attention networks and the intrinsic network structure of the human brain. *Human Brain Mapping, 43*, 1431–1448. https://doi.org/10.1002/hbm.25734

Martinussen, R., Hayden, J. H.-J., & Tannock, R. (2005). A meta-analysis of working memory impairments in children with attention-deficit/hyperactivity disorder. *Journal of the American Academy of Child & Adolescent Psychiatry, 44*, 377–384. https://doi.org/10.1097/01.chi.0000153228.72591.73

Mattfield, A., Whitfield-Gabrieli, S., Biederman, J., Spencer, T., Brown, A., et al. (2016). Dissociation of working memory impairments and attention-deficit/hyperactivity disorder in the brain. *NeuroImage: Clinical, 10*, 274–282. https://doi.org/10.1016/j.nicl.2015.12.003

Mayes, S., Calhoun, S., Mayes, R., & Molitoris, S. (2012). Autism and ADHD: Overlapping and discriminating symptoms. *Research in Autism Spectrum Disorders, 6*(1), 277–285. https://doi.org/10.1016/j.rasd.2011.05.009

McCormick, E., Arnemann, K., Ito, T., Hanson, S., & Cole, M. (2022). Latent functional connectivity underlying multiple brain states. *Network Neuroscience, 6*(2), 570–590. https://doi.org/10.1162/netn_a_00234

McLennan, J. (2016). Understanding attention deficit hyperactivity disorder as a continuum. *The Official Journal of the College of Family Physicians Canada, 62*(12), 979–982.

Meek, W., & Benson, A. (2002). Dissecting the Brain's internal clock: How frontal–striatal circuitry keeps time and shifts attention. *Brain and Cognition, 48*(1), 195–211. https://doi.org/10.1006/brcg.2001.1313

Melloni, L., van Leeuwen, S., Alink, A., & Muller, N. (2012). Interaction between bottom-up saliency and top-down control: How saliency maps are created in the human brain. *Cerebral Cortex, 22*(12), 2943–2052. https://doi.org/10.1093/cercor/bhr384

Menon, V. (2015). *Brain mapping: An encyclopedic reference* (Vol. 2). Academic.

Menon, V., & Uddin, L. (2010). Saliency, switching, attention and control: A network model of insula function. *Brain Structure and Function, 214*(5–6), 655–667. https://doi.org/10.1007/s00429-010-0262-0

Meyer, K., & Damasio, A. (2009). Convergence and divergence in a architecture for recognition and memory. *Trends in Neuroscience, 32*(7), 376–382. https://doi.org/10.1016/j.tins.2009.04.002

Milich, R. B., & Lynam, D. (2001). ADHD combined type and ADHD predominantly inattentive type are distinct and unrelated disorders. *Clinical Psychology: Science and Practice, 8*(4), 463–488. https://doi.org/10.1093/clipsy.8.4.463

Mitchell-Hegg, R., Prado, S., Gava, G., Go, M., & Schultz, S. (2023). Neural manifold analysis of brain circuit dynamics in health and disease. *Journal of Consultational Neuroscience, 51*(1), 1–21. https://doi.org/10.1007/s10827-022-00839-3

Mohr, K., Carr, N., Georgel, R., & Kelly, S. (2020). Modulation of the earliest component of the human VEP by spatial attention: An investigation of task demands. *Cerebral Cortex Communications, 1*(1), tgaa045.
Molloy, M. S. (2022). Individual variability in functional organization of the neonatal brain. *NeuroImage, 253*, 119101. https://doi.org/10.1016/j.neuroimage.2022.119101
Monk, C. S., Klein, R. G., Telzer, E. H., Schroth, E. A., Mannuzza, S., Moulton, J. L., 3rd, Guardino, M., Masten, C. L., McClure-Tone, E. B., Fromm, S., Blair, R. J., Pine, D. S., & Ernst, M. (2008). Amygdala and nucleus accumbens activation to emotional facial expressions in children and adolescents at risk for major depression. *American Journal of Psychiatry, 165*(1), 90–98.
Mulder, H., Pitchford, N., Hagger, M., & Marlow, N. (2009). Development of executive function and attention in preterm children: A systematic review. *Developmental Neuropsychology, 34*(4), 393–421. https://doi.org/10.1080/87565640902964524
Muller, J., & Roberts, J. (2005). Memory and attention in obsessive–compulsive disorder: A review. *Journal of Anxiety Disorders, 19*(1), 1–26. https://doi.org/10.1016/j.janxdis.2003.12.001
Murty, V., & Adcock, R. (2014). Enriched encoding: Reward motivation organizes cortical networks for hippocampal detection of unexpected events. *Cerebral Cortex, 24*(8), 2160–2168.
Nazari, R., & Salehi, M. (2023). Early development of the functional brain network in newborns. *Brain Structure & Function, 228*, 1725–1739. https://doi.org/10.1007/s00429-023-02681-4
Nelson, A., & Bender, K. (2021). Dendritic integration dysfunction in neurodevelopmental disorders. *Developmental Neuroscience, 43*(3–4), 201–221. https://doi.org/10.1159/000516657
Newell, A. (1956). The logic theory machine: A complex information processing system. *IRE Transactions on Information Theory, 2*(3), 6179. https://doi.org/10.1109/TIT.1956.1056797
Oberauer, K. (2019). Working memory and attention—A conceptual analysis and review. *Journal of Cognition, 2*, 1. (36). https://doi.org/10.5334/joc.58
Oldham, S., & Fornito, A. (2019). The development of brain network hubs. *Developmental Cognitive Neuroscience, 36*, 100607.
Oldham, S., Fulcher, B., Parkes, L., Arnatkevic̆iūtė, A., Suo, C., & Fornito, A. (2019). Consistency and differences between centrality measures across distinct classes of networks. *PLoS One, 14*(7), e0220061. https://doi.org/10.1371/journal.pone.0220061
Onofrj, V., Chiarelli, A., Wise, R. C., & Caulo, M. (2022). Interaction of the salience network, ventral attention network, dorsal attention network and default mode network in neonates and early development of the bottom-up attention system. *Brain Structure and Function, 227*, 1843. https://doi.org/10.1007/s00429-022-02477-y
Pace-Schott, E. P. (2017). *In principles and practice of sleep medicine*. Elsevier.
Pardo, J. V., Pardo, P. J., & Janer, K. R. (1990). The anterior congulate cortex, mediates processing selection in the Stroop attentional conflict paradigm. *Proceedings. National Academy of Sciences. United States of America, 86*(1), 256–259.
Park, H., & Friston, K. (2013). Structural and functional brain networks: From connections to cognition. *Science, 342*(6158), 1238411.
Parr, T., & Friston, K. (2019). Attention or salience? *Current Opinion in Psychology, 29*, 1–5. https://doi.org/10.1016/j.copsyc.2018.10.006
Pavlov, I. &. (1927). *Conditioned reflexes: An investigation of the physiological activity of the cerebral cortex*. Oxford University Press.
Peelan, M., Heslenfeld, D., & Theeuwes, J. (2004). Endogenous and exogenous attention shifts are mediated by the same large-scale neural network. *NeuroImage, 22*(2), 822–830. https://doi.org/10.1016/j.neuroimage.2004.01.044
Penrose, O., & Liebowitz, J. (1979). Toward a rigorous molecular theory of metastability. In E. Montroll & J. Leibowitz (Eds.), *Studies in statisical mechanics II, fluctuation phenomena* (pp. 319–322). North Holland Publishing Company.
Petersen, S., & Posner, M. (2012). The attention system of the human brain: 20 years after. *Annual Review of Neuroscience, 35*, 73–89. https://doi.org/10.1146/annurev-neuro-062111-150525. PMID: 22524787; PMCID: PMC3413263.

Petersen, S., Fox, P., Miezin, F., & Raichle, M. (1988). *Modulation of cortical visual responses by direction of spatial attention measured by PET* (p. 22). Association for Research in Vision Ophthalmology.

Pharisi, D., & Nolfi, F. (2001). *Development in neural networks*. Retrieved from MIT Press Direct: https://direct.mit.edu/books/edited-volume/2039/chapter-abstract/55299/Development-in-Neural-Network

Pinto, Y., van der Leij, A., Sligte, S., Lamme, V., & Scholte, S. (2013). Bottom-up and top-down attention are independent. *Journal of Vision, 13*, 16. https://doi.org/10.1167/13.3.16

Posner, M., & Petersen, S. (1990). The attention system of the human brain. *Annual Review of Neuroscience, 13*(1), 25–42. https://doi.org/10.1146/annurev.ne.13.030190.000325

Posner, J., Polanczyk, G., & Sonuga-Barke, E. (2020). Attention-deficit hyperactivity disorder. *Lancet, 395*(10), 450–462. https://doi.org/10.1016/S0140-6736(19)33004-1

Raichle, M. (2015). The brain's default mode network. *Annual Review of Neuroscience, 38*(July), 433–447. https://doi.org/10.1146/ANNUREV-NEURO-071013-014030

Raiker, J. S., Friedman, L. M., Orban, S. A., Kofler, M. J., Sarver, D. E., & Rapport, M. D. (2019). Phonological working memory deficits in ADHD revisited: The role of lower level information-processing deficits in impaired working memory performance. *Journal of Attention Disorders, 23*(6), 570–583. https://doi.org/10.1177/1087054716686182

Rapport, M., Bolden, J., Kofler, M., Sarver, D., Raiker, J., & Alderson, R. (2009). Hyperactivity in boys with attention-deficit/hyperactivity disorder (ADHD): A ubiquitous core symptom or manifestation of working memory deficits? *Journal of Abnormal Child Psychology, 37*, 521–534.

Raye, C., Johnson, M., Mitchell, E., Greene, K., & Johnson, M. R. (2007). Refreshing: A minimal executive function. *Cortex, 43*, 145–145. https://doi.org/10.1016/S0010-9452(08)70451-9

Reimherr, F., Roesler, M., Marchant, B., Gift, T., Tetz, W., & Philipp-Wiegmann, F. (2020). Types of adult attention-deficit/hyperactivity disorder: A replication analysis. *The Journal of Clinical Psychiatry, 81*, 2. https://doi.org/10.4088/JCP.19m13077

Roberts, C. S. (2023). Impact and centrality of attention dysregulation on cognition, anxiety and low mood in adolescents. *Scientific Reports, 13*, 9106.

Roda, C., & Thomas, J. (2006). Attention aware systems: Theories, applications, and research agenda. *Computers in Human Behavior, 22*, 557–587.

Rolls, E., & Stringer, S. (2000). On the design of neural networks in the brain by genetic evolution. *Progress in Neurobiology, 61*(6), 557–579. https://doi.org/10.1016/S0301-0082(99)00066-0

Rolls, E., McCabe, C., & Redoute, J. (2008). Expected value, reward outcome, and temporal difference error representations in a probabilistic decision task. *Cerebral Cortex, 18*(3), 652–663.

Rosales, A., Vitoratou, S., Banaschewski, T., Asherson, P., Buitelaar, J., Oades, R., et al. (2015). Are all the 18 DSM-IV and DSM-5 criteria equally useful for diagnosing ADHD and predicting comorbid conduct problems? *European Child Adolescent Psychiatry, 24*, 1325–1337. https://doi.org/10.1007/s00787-015-0683-7

Rost, T., Deger, M., & Nawrot, M. (2018). Winnerless competition in clustered balanced networks: Inhibitory assemblies do the trick. *Biological Cybernetics, 112*(5283), 81. https://doi.org/10.1007/s00422-017-0737-7

Saad, J., et al. (2022). Intrinsic functional connectivity in the default mode network differentiates the combined and inattentive attention deficit hyperactivity disorder types. *Frontiers in Human Neuroscience, 16*. https://doi.org/10.3389/fnhum.2022.859538

Saboory, E., Ghasemi, M., & Mehranfard, D. (2020). Norepinephrine, neurodevelopment and behavior. *Neurochemistry International, 135*, 104706. https://doi.org/10.1016/j.neuint.2020.104706

Salehinejad, M., Ghayerin, E., Nejati, V., & Yavari, F. N. (2020). Domain-specific involvement of the right posterior parietal cortex in attention network and attentional control of ADHD: A randomized, cross-over, sham-controlled tDCS study. *Neuroscience, 444*, 149–159. https://doi.org/10.1016/j.neuroscience.2020.07.037

Sali, A., Aderson, B., & Yantis, S. (2014). The role of reward prediction in the control of attention. *Journal of Experimental Psychology: Human Perception and Performance, 40*(4), 1654–1664. https://doi.org/10.1037/a0037267

Salmi, S., Rinne, T., Koistien, S., Salonen, O., & Alho, K. (2009). Brain networks of bottom-up triggered and top-down controlled shifting of auditory attention. *Brain Research, 1286*, 155–164. https://doi.org/10.1016/j.brainres.2009.06.083

Salmi, J., Salmela, V., Salmo, E., Mikkola, K., Leppämäki, S., Tani, P., & Hokkanen, H. (2018). Out of focus – Brain attention control deficits in adult ADHD. *Brain Research, 1692*, 12–22. https://doi.org/10.1016/j.brainres.2018.04.019

Sani, I., Stemmann, H., Caron, B., & al., e. (2021). The human endogenous attentional control network includes a ventro-temporal cortical node. *Nature Communications, 12*, 360. https://doi.org/10.1038/s41467-020-20583-5

Sato, T., Watanabe, K., Thompson, K., & Schall, J. (2003). Effect of target-distractor similarity on FEF visual selection in the absence of the target. *Experimental Brain Research, 151*(3), 356–363.

Scahachar, R. (2023). Fifty years of executive control research in attention-deficit/hyperactivity disorder: What we have learned and still need to know. *Neuroscience & Biobehavioral Reviews, 155*, 105461. https://doi.org/10.1016/j.neubiorev.2023.105461

Schimmelpfennig, J., Topczewski, J., Zajkowski, W., & Jankowiak-Siuda, K. (2023). The role of the salience network in cognitive and affective deficits. *Frontiers in Human Neuroscience., 17*, 1133367. https://doi.org/10.3389/fnhum.2023.1133367

Schneidt, S., Jusyte, A., Rauss, K., & Schönenberg, M. (2018). Distraction by salient stimuli in adults with attention-deficit/hyperactivity disorder: Evidence for the role of task difficulty in bottom-up and top-down processing. *Cortex, 101*, 206–220. https://doi.org/10.1016/j.cortex.2018.01.021

Schweitzer, J., Hanford, R., & Medoff, D. (2006). Working memory deficits in adults with ADHD: Is there evidence for subtype differences? *Brain and Behavioral Functions, 2*(43). https://doi.org/10.1186/1744-9081

Science Direct. (2020). *Precuneus*. Retrieved from Science Direct: https://www.sciencedirect.com/topics/neuroscience/precuneus#:~:text=The%20precuneus%20is%20a%20brain,and%20affective%20responses%20to%20pain

Scott, S. H., & Kalaska, J. F. (1997). Reaching movements with similar hand paths but different arm orientations. I. Activity of individual cells in motor cortex. *Journal of Neurophysiology, 77*, 826–852.

Seeley, W. (2019). The salience network: A neural system for perceiving and responding to homeostatic demands. *Journal of Neuroscience, 39*(50), 9878–9882. https://doi.org/10.1523/JNEUROSCI.1138-17.2019

Shen, K., et al. (2012). Information processing architecture of functionally defined clusters in the macaque cortex. *Journal of Neuroscience, 32*(48), 17465–17476. Epub 2012/12/01.

Singer, W. (1993). Synchronization of cortical activity and its putative. *Annual Review of Physiology, 55*, 349–374.

Skirrow, C., & Asherson, P. (2013). Emotional lability, comorbidity and impairment in adults with attention-deficit hyperactivity disorder. *Journal of Affective Disorders, 147*(1–3), 273–281.

Sohn, H., Meirhaeghe, N., Rajalingham, R., & Jazayeri, M. (2021). A network perspective on sensorimotor learning. *Trends in Neuroscience, 44*(3), 170–181. https://doi.org/10.1016/j.tins.2020.11.007

Song, H., Shim, W., & Rosenebrg, M. (2023). Large-scale neural dynamics in a shared low-dimensional state space reflect cognitive and attentional dynamics. *Neuroscience*. https://doi.org/10.7554/eLife.85487

Sonuga-Barke, E. (2005). Causal models of attention-deficit/hyperactivity disorder: From common simple deficits to multiple developmental pathways. *Biological Psychiatry, 57*, 1231–1238.

Stam, C. (2024). Hub overload and failure as a final common pathway in neurological brain network disorders. *Network Neuroscience, 8*(1), 1–23. https://doi.org/10.1162/netn_a_00339

Sterzer, P., Voss, M., Schlagenhauf, F., & Heinz, A. (2019). Decision-making in schizophrenia: A predictive-coding perspective. *NeuroImage, 190*, 133–143. https://doi.org/10.1016/j.neuroimage.2018.05.074

Sturm, W., & Willmes, K. (2001). On the functional neuroanatomy of intrinsic and phasic alertness. *NeuroImage, 14*, 76–84.

Stuss, D. (2006). Frontal lobes and attention: Processes and networks, fractionation and integration. *Journal of the International Neuropsychological Society, 12*(2), 261–271. https://doi.org/10.1017/S1355617706060358

Suarez, L., Markello, R., Betzel, R., & Misic, B. (2020). Linking structure and function in macroscale brain networks. *Trends Cognitive Science, 24*(4), 302–315. Epub 2020/03/12. pmid:32160567.

Suslow, T., Hußlack, A., Kersting, A., & Bodenschatz, C. M. (2020). Attentional biases to emotional information in clinical depression: A systematic and meta-analytic review of eye tracking findings. *Journal of Affective Disorders*, 632–642. https://doi.org/10.1016/j.jad.2020.05.140

Tegelbeckers, J., Bunzeck, N., Duzel, E., Fletchtner, H., & Krauel, K. (2015). Altered salience processing in attention deficit hyperactivity disorder. *Human Brain Mapping, 36*, 2049–2060. https://doi.org/10.1002/hbm.22755

Thapar, A., Langley, K., O'Donovan, M., & Owen, M. (2006). Refining the attention deficit hyperactivity disorder phenotype for molecular genetic studies. *Molecular Psychiatry, 11*(8), 714–720.

Theeuwes, J. (2018). Visual selection: Usually fast and automatic; seldom slow and volitional. *Journal of Cognition, 1*(1), 29. https://doi.org/10.5334/joc.13

Tognoli, E., & Kelso, J. (2013). The metastable brain. *Neuron, 81*, 35. https://doi.org/10.1016/j.neuron.2013.12.022

Treisman, A. (1969). Strategies and models of selected attention. *Psychological Review, 76*, 282–299.

Treisman, A., & Gelade, G. (1980). A feature-integration theory of attention. *Cognitive Psychology, 12*, 97–136.

Tsuchida, A., & Fellows, L. (2013). Are core component processes of executive function dissociable within the frontal lobes? Evidence from humans with focal prefrontal damage. *Cortex, 49*(7), 1790–1800. https://doi.org/10.1016/j.cortex.2012.10.014

Tsujimoto, S. (2008). The prefrontal cortex: Functional neural development during early childhood. *The Neuroscientist, 14*, 345–358.

Tyukin, I., Gorban, A., Calvo, C., Makarova, J., & Makarov, V. (2018). High-dimensional brain: A tool for encoding and rapid learning of memories by single neurons. *Bulletin of Mathematical Biology, 81*(11), 4856–4888. https://doi.org/10.1007/s11538-018-0415-5

Ungerleider, L., & Haxby, J. (1994). What' and 'where' in the human brain. *Current Opinion in Neurobiology, 4*, 157–165.

van den Heuvel, M., & Sporns, O. (2013). Network hubs in the human brain. *Trends in Cognitive Sciences, 17*(12), 683–696. https://doi.org/10.1016/j.tics.2013.09.012

van den Heuvel, M., et al. (2018). Hubs in the human fetal brain network. *Developmntal Cognitive Neuroscience, 30*, 108. https://doi.org/10.1016/j.dcn.2018.02.001

van der Meer, J., Oerlemans, A., van Steijn, D. J., Lappenschaar, M. G., de Sonneville, L. M., Buitelaar, J. K., & Rommelse, N. N. (2012). Are autism spectrum disorder and attention-deficit/hyperactivity disorder different manifestations of one overarching disorder? Cognitive and symptom evidence from a clinical and population-based sample. *Journal American Academy Child Adolescent Psychiatry, 51*(11), 1160–1172. https://doi.org/10.1016/j.jaac.2012.08.024

Vanunu, Y., Hotaling, J., Le Pelley, M., & Newell, B. (2021). How top-down and bottom-up attention modulate risky choice. *Proceedings. National Academy of Sciences. United States of America, 118*(39), 10.1073/pnas.2025646118.

Visintin, E., De Panfilis, C., Antonucci, C., Capecci, C., Marchesi, C., & Sambataro, F. (2015). Parsing the intrinsic networks underlying attention: A resting state study. *Behavioural Brain Research, 278*, 315–322. https://doi.org/10.1016/j.bbr.2014.10.002

Vossel, S., Geng, J., & Fink, G. (2014). Dorsal and ventral attention systems: Distinct neural circuits but collaborative roles. *The Neuroscientist, 20*(2), 150–159. https://doi.org/10.1177/1073858413494269

Wang, Y. M. (2024). The neural and genetic underpinnings of different developmental trajectories of attention-deficit/hyperactivity symptoms in children and adolescent. *BMC Medicine, 22*. https://doi.org/10.1186/s12916-024-03449-1

Wang, B., & Theeuwes, J. (2020). Salience determines attentional orienting in visual selection. *Journal of Experimental Psychology: Human Perception and Performance, 46*(10), 1051–1057. https://doi.org/10.1037/xhp0000796

Wang, W., Shen, J., Cheng, M., & Shao, L. (2019). An iterative and cooperative top-down and bottom-up inference network for salient object detection. In *Proceedings of the IEEE/CVF conference on computer vision and pattern recognition* (pp. 5968–5977). CVF.

Wasserman, T., & Wasserman, L. (2013). Toward an integrated model of executive functioning in children. *Journal of Applied Neuropsychology: Child, 2*(2), 88–96. https://doi.org/10.1080/21622965.2013.748394

Wasserman, T., & Wasserman, L. (2015). The misnomer of attention deficit hyperactivity disorder. *Applied Neuropsychology: Child*, 1–7. https://doi.org/10.1080/21622965.2015.1005487

Wasserman, T., & Wasserman, L. (2020). *Motivation, effort and the neural network model*. Springer Nature.

Wassserman, L. W. (2024). *Motor network and implications for neurodevelopmental disorders*. Costa Mesa.

Weber-Goericke, F., & Muehlhan, M. (2019). A quantitative meta-analysis of fMRI studies investigating emotional processing in excessive worriers: Application of activation likelihood estimation analysis. *Journal of Affective Disorders, 243*, 348–359. https://doi.org/10.1016/j.jad.2018.09.04

Wen, X., et al. (2019). First-year development of modules and hubs in infant brain functional networks. *NeuroImage, 185*, 222–235. https://doi.org/10.1016/j.neuroimage.2018.10.019

Willcut, E., Doyle, A., Nigg, J., Faraone, S., & Pennington, B. (2005). Validity of the executive function theory of attention-deficit/hyperactivity disorder: A meta-analytic review. *Biological Psychiatry, 57*(11), 1336–1346.

Wolfe, J., Cave, K., & Franzel, S. (1989). Guided search: An alternative to the feature integration model for visual search. *Journal Experimental Psychology: Human Perception and Performance, 15*(3), 419–433.

Woodman, G., Luck, S., & Schall, J. (2007). The role of working memory representations in the control of attention. *Cerebral Cortex., 17*, i118–i124. https://doi.org/10.1093/cercor/bhm065

Xuan, B., Mackie, M., Spagna, A., Wu, T., Tian, Y., Hof, P., & Fan, J. (2016). The activation of interactive attentional networks. *NeuroImage*, 308–319. https://doi.org/10.1016/j.neuroimage.2016.01.017

Yang, W., Ding, Z., Dai, T., Peng, F., & Zhang, J. (2015). Attention bias modification training in individuals with depressive symptoms: A randomized controlled trial. *Journal of Behavior Therapy and Experimental Psychiatry, 49*(Part A), 101–111. https://doi.org/10.1016/j.jbtep.2014.08.005

Yao, Y., Xuan, Y., Wu, R., & Sand, B. (2020). Regulatory effects of reward anticipation and target on attention processing of emotional stimulation. *Fronteirs on Psychology*. https://doi.org/10.3389/fpsyg.2020.01170

Yiend, J. (2010). The effects of emotion on attention: A review of attentional processing of emotional information. In I. J. de Houwer & D. Hermans (Eds.), *Cognition and emotiona: Current research and theories*. Psychology Press.

Zang, Y., Li, Q., Wang, Z., Liu, X., & Zheng, Y. (2017). Temporal dynamics of reward anticipation in the human brain. *Biological Psychology, 128*, 89–97. https://doi.org/10.1016/j.biopsycho.2017.07.011

Zhang, Y. C., Xin, Y., Peng, B., & Liu, S. (2023). Norepinephrine system at the interface of attention and reward. *Progress in Neuro-Psychopharmacology and Biological Psychiatry, 125*, 110751. https://doi.org/10.1016/j.pnpbp.2023.110751

Zhao, T., Mishra, V., et al. (2019). Structural network maturation of the preterm human brain. *NeuroImage, 185*, 699–710. https://doi.org/10.1016/j.neuroimage.2018.06.047

Zhao, Y., Yang, L., Gong, G., Cao, Q., & Liu, J. (2022). Identify aberrant white matter microstructure in ASD, ADHD and other neurodevelopmental disorders: A meta-analysis of diffusion tensor imaging studies. *Progress in Neuro-Psychopharmacology and Biological Psychiatry, 113*, 110477. https://doi.org/10.1016/j.pnpbp.2021.110477

Zohary, E., Shadlen, M., & Newsome, W. (1994). Correlated neuronal discharge rate and its implications for psychophysical performance. *Nature, 370*, 140–143.

Zorcec, T., & Jordanova, N. (2010). ADHD as an executive dysfunction. *Contributions, Section Biological, Medical Science, 31*(2), 171–181.

Zuo, N., Yang, Z., Li, J., & Jiang, T. (2018). Core networks and their reconfiguration patterns across cognitive loads. *Human Brain Mapping, 39*(9), 3546–3557. https://doi.org/10.1002/hbm.24193

Chapter 11
Emotional Dysregulation and Attention

Emotional regulation has been described as an individual's ability to modify an emotional state so as to promote adaptive, goal-oriented behaviors (Thompson, 1994). It describes the processes that permit an individual to select, attend to, and evaluate emotionally arousing stimuli in a flexible manner. These processes result in behavioral and physiological responses that can be modulated in line with value-determined goals. Emotional dysregulation is what happens when these adaptive processes are impaired or disrupted. This impairment leads to behavior that is often not in the individual's best interests. Emotional dysregulation incorporates (Thompson, 1994): (1) emotional expressions and experiences that are excessive in relation to social norms and are context inappropriate; (2) rapid, poorly controlled shifts in emotion (lability); and (3) the abnormal allocation of attention to emotional stimuli. Clearly, given the focus of this book, it is this last aspect of emotional dysregulation that will be discussed here.

In this chapter we will review the interaction of emotional dysregulation, attention, and specific clinical conditions in detail. In order to address the future content of this chapter, let us review the cortical and subcortical relationships to domains of current interest. The regions of the prefrontal cortex (PFC) that have most consistently been implicated in cognitive control processes, including decision-making, salience, and emotional regulation, include the orbitofrontal cortex (OFC), dorsomedial prefrontal cortex (MdPFC), anterior cingulate gyrus (ACG), dorsolateral prefrontal cortex (DLPFC), and ventrolateral prefrontal cortex (VLPFC)) (Phillips et al., 2008). The anterior cingulate cortex (ACC) has connections to both the "emotional" or limbic system and the "cognitive" or prefrontal cortex. Functional connectivity research has demonstrated a dorsal–rostral division in ACC modulation of the thalamus. This suggests an important role in the integration of neuronal circuitry for affect regulation. This part of the sensory cortex pathway responds to emotionally salient stimuli, with the dorsal ACC showing a positive modulation of this pathway, and the rostral ACC, a negative modulation. Collectively, these findings

T. Wasserman, L. D. Wasserman, *Disorders of Attention*, Neural Network Model: Applications and Implications, https://doi.org/10.1007/978-3-031-78732-4_11

therefore implicate different regions of the PFC, including OFC, MdPFC, VLPFC, and ACG, and further suggest distinguishable roles of these regions in the regulation of activity within subcortical limbic regions in the response to emotionally salient information. This then involves emotional components, motivation, reward, and salience. It should be pointed out that salience is a key subprocess of attention control (allocation).

Attention and Emotional Dysregulation

As the above section highlights, attention and emotion are highly integrated. Our decision, our motivation, and our appreciation of rewards depend upon this synchrony. What happens when this interplay becomes dysregulated? Attention dysregulation has been referred to as inefficient regulation or control of self-motivational systems directing normative top-down goal-directed action (Roberts, 2023). Attention dysregulation is a core characteristic of attention deficit hyperactivity disorder (ADHD). It is important to note that ADHD is a disorder that usually developmentally precedes clinical anxiety and mood disorders (Solmi et al. 2021). The presence of ADHD in a developmental profile appears to increase their future likelihood of anxiety and/or other mood disorders emerging. Please note that the former author refers to the trajectory leading to a disorder. The current authors wish to highlight their position that the processes impacting emotional regulation begin at birth and are in fact part of the overall foundational developmental system (Wassserman, 2024). We would argue that internal arousal, environmental arousal, and their ties to emotional regulation are already interconnected in the neonate. Appropriate alerting, arousal, and emotional regulation are portents to healthy neurodevelopment. Poor integration may be presented later on in many disorders. The attention component is then associative, not necessarily causative. That is, ADHD does not *cause* anxiety or depressive disorders but is rather part of an initially poorly regulated system with attention, initially alerting, and arousal, playing a major role, and a developmental marker for potentially trajectory-based development of later unfolding disorders.

Emotional regulation is the ability to recognize, evaluate, modify, and manage emotions in a personal and socially acceptable way in order to maintain mental control over strong feelings and arrive at adaptive functioning. In contrast, emotional dysregulation (ED) is a transdiagnostic construct defined as the inability to regulate the intensity and quality of emotions (such as fear, anger, sadness) in order to generate an appropriate emotional response, to handle excitability, mood instability, and emotional overactivity, and to return to an emotional baseline (Paulus et al., 2021). The transdiagnostic nature of the construct speaks to emotional dysregulation appearing in many diagnoses. Therefore, while of significant importance, it is not always the primary nor necessarily dissociable factor. Nonetheless, the implications of emotional dysregulation can be quite damaging.

While many authors describe the developmental process of infants, for the current authors the developmental perspective begins at birth. For us, and others, the rudimentary ability to regulate emotions and the process to developing emotional regulation begins at birth; the neonate responds to internal cues (hunger) and expresses emotional distress. An appropriate response from a caregiver sets the stage for healthy integration of emotional regulation (Wassserman, 2024). Infants go on to learn from interacting with their caregivers and their environment to differentiate their primary emotional states (Gross, 2013). They also learn that these states are variable in intensity, and that this intensity can be modified later on by using self-control, self-soothing, or distracting. Several processes, such as the development of executive functions and language, influence the development of a child regulatory skills (Groves et al., 2020). Emotional regulation also involves the accessing of coping resources, using a broad range of regulation strategies. The development of these strategies is complex, involving genetics, epigenetics, cognition, social experiences, and learning (Ross & Thompson, 2011; Marquis et al., 2017). In fact, differences in emotional regulation, including self-soothing, can be observed in the neonate (Brazelton, 1973). When this process goes awry, emotional control does not develop correctly and is dysregulated. For example, a longitudinal study of over 7000 children found that temperamental emotionality at age 3 predicted comorbid ADHD at age 7 (Stringaris et al., 2010). The current authors would consider this an important example of how important the consideration of developmental trajectories is. Again, this emotionality/ADHD association is correlated, not causative.

In fact, the co-occurrence of attention dysregulation is associated with a wide range of adverse outcomes related to emotional dysregulation, psychopathological features, and executive functioning (EF) impairments (Roberts, 2023). Their research found important bridging traits and socioenvironmental influences in the relationships between attention dysregulation symptoms related to ADHD and mood/anxiety disorders. Trait perfectionism was uniquely associated with both better cognitive performance and broad psychopathological dimensions. The results suggested that attentional dysregulation may moderate the breadth of EF. That is, the level of attention regulation may moderate fluid and crystalized cognitive task outcomes seen in adolescents with anxiety and low mood and may be central to various pathological features. As a result, this might provide a target for reducing wide-ranging negative developmental outcomes in a number of conditions where emotions are dysregulated.

Further, the relationship of attentional dysregulation to emotional dysregulation may not be as has been traditionally conceived. It is well known that psychopathology is often associated with, and predictive of, cognitive impairments, especially attentional deficiencies (Brieant et al., 2022). It is of course quite logical to assume that the psychopathology caused the attention controls of the individual to be aberrant. This might not, however, be the case. The results of Roberts (2023) suggest that underperformance in a wide range of cognitive tasks inclusive of the regulation of attention is due to attentional dysregulation, especially in individuals experiencing anxiety or low mood (and perhaps depression in some cases). Their research indicated that *overall executive function, inclusive of attentional control*, is likely a predictor of change in the general psychopathological status of the individual and that these deficiencies may, over time, manifest into psychopathology (Romer & Pizzagalli, 2021). This finding was also supported by the work of Wehmeier et al. (2010), who found that individuals with both ADHD and emotion dysregulation were significantly more impacted in the areas of peer relationships, family life, occupational attainment, and academic performance than those individuals diagnosed with ADHD alone. We surmise that this should not come as a surprise for several reasons. These findings can represent both an increase in the presentation of the number of issues as well as the magnitude of the effect. Notwithstanding the effects of dealing with attention issues, the lack of predictability for both the emotional responsivity in an excessive or underresponsive way can easily be imagined to impact personal, academic, and occupational relationships in magnitude. Going on to explore executive function makes this point of magnitude of effect for the interplay of attention and emotion regulation even more strongly.

Attention and Anxiety

Attention dysregulation is a robust factor underlying wide-ranging cognitive task impairments seen in adolescents with moderate to severe anxiety. Attentional control theory (ACT) (Eysenck & Derakshan, 2011) proposes that attention control (AC) deficits are central to the development of anxiety. Derakshan and Eysenck's (2009) attentional control theory posits that anxiety impairs the efficiency of specific aspects of executive functioning, inhibition, and shifting. A second core theoretical assumption is that anxiety impairs the *quality* of executive performance to a lesser extent than the *efficiency* of executive performance. In this model, quality refers to outcome measures. Processing efficiency represents the relationship between performance effectiveness and effort, or use of processing resources such as working memory.

In a meta-analysis evaluating the accuracy of these assumptions, Shi et al. (2019) found that their results supported hypothesis of ACT. Specifically, anxiety produced significant deficits in attention control efficiency, but not effectiveness. These significant deficits occurred in inhibition and switching, but not updating. Studies within the meta-analytic sample having high cognitive load conditions found larger anxiety-related attention control deficits. Age moderated the relationship between anxiety and attention control in behavioral studies and anxiety severity moderated this relationship in self-report studies. The negative impact of anxiety on these specific executive functions appears to increase with age.

Historically, the predominant view of anxiety (Derakshan & Eysenck, 2009) defines it as a motivational state that exists when there is a high level of perceived threat to that individual (Derakshan & Eysenck, 2009). ACT posits that anxiety impacts the balance between regulating stimulus-driven (bottom-up processing) and goal-driven (top-down) processing with greater attention paid to the bottom-up stimulus-driven processing over the more efficient goal-driven process. They point out that the worry and use of strategies to mitigate the effects of the anxiety will then result in adverse effects on cognitive performance. This stance has research support (Pergamin-Hight et al., 2015). Under the perception of threat, state anxiety impairs cognitive performance by negatively impacting processes such as working memory and allocation of attentional resources. In addition, research into attentional bias in children has found that individuals high in either state or trait anxiety tend to preferentially allocate their attention to threatening stimuli compared to neutral stimuli, and that this bias is greater than observed in non-anxious individuals or under low anxiety conditions (Dudeney et al., 2015). They also conclude that the difference increases with age.

The contribution of attentional bias to the development of anxiety would suggest that addressing the attentional bias would decrease the symptoms of anxiety in general and also the anxiety associated with specific clinical conditions. There is support for this position. Kuckertz et al. (2014) noted that individuals with post-traumatic stress disorder possess an attentional bias toward threat-relevant information. They hypothesized that attention bias modification (ABM), an effective treatment for anxiety disorders (Beard et al., 2012), would be efficacious in reducing the anxiety in post-traumatic stress disorder and reduce the clinical impact of post-traumatic stress disorder on affected individuals. The results of Sawyer et al. indicated attention bias across comparison conditions, whereas appetitive stimuli (pleasant or wanted stimuli that an organism will approach or voluntarily contact) produced changes in attention only when comparing appetitive versus neutral conditions. In summary, ABM resulted in a moderate and robust effect on attention bias when using threat stimuli, and reduced the clinical impact of the post-traumatic stress disorder on the functioning of the impacted individual. Here is an example of a treatment for attention inefficiency that is not related to a customary treatment for attention deficit disorder or to a historical treatment of post-traumatic stress disorder that produces disorder-specific positive effects. Similar effects of targeting attentional biases with specific attentional bias reduction protocols have been reported in studies of social anxiety (Carleton et al., 2015).

Some neuroimaging is now focusing on functional networks and anxiety as a means of differentiating anxiety disorders from other disorders, for example a negative mood state such as depression. In fact, there is research to suggest a specific pattern of white matter connectivity impacting attention that is associated with anxiety disorders (Sylvester et al., 2012). Their research found that anxiety disorders and high trait anxiety are associated with a particular pattern of functional network dysfunction: increased functioning of the cingulo-opercular and ventral attention networks as well as decreased functioning of the frontoparietal and default mode networks. They posited that using a functional network model could differentiate the attentional pathology of anxiety disorders from other psychiatric illnesses such as major depression and provides targets for novel treatment strategies. On a larger scale, they posit that viewing the functional networks from a dimensional perspective in which each functional network can be viewed as underactive, normal, or overactive, with different "blends" of disturbance along these dimensions, might allow for a network-wide, rather than symptom-based, understanding of psychiatric disorders.

Imperatori et al. (2019) looked at trait anxiety as a relatively stable factor and high anxiety as a dimensional trait of anxiety. They defined anxiety trait as a crucial vulnerability factor, particularly for mood and anxiety disorders. Their particular focus was to investigate the default mode network (DMN) connectivity, an area frequently associated with psychopathology, in individuals with high trait anxiety using quantitative electroencephalography (EEG). They found that compared to controls, individuals with high-trait-anxiety showed a decrease of theta connectivity between right medial prefrontal cortex (mPFC) and right posterior cingulate/retrosplenial cortex. A decrease of beta connectivity was also observed between right mPFC and right anterior cingulate cortex. In addition, the strength of DMN functional connectivity suggested that lower connectivity was associated with higher trait anxiety. This result held true even when controlling for potential confounding variables such as sex, age, and general psychopathology. They posited that high-trait-anxiety individuals fail to synchronize DMN during resting states and that this reflected a possible top-down cognitive control deficit in individuals with trait anxiety.

As demonstrated, anxiety significantly impacts the allocation of attention. When it works correctly, it assists the individual in focusing on potential threats in the environment. When it works inefficiently, it focusses the individual on suboptimal stimuli in the environment. How can we understand how anxiety can be both a positive and negative factor at the same time? One way to understand this phenomenon is by utilizing signal detection theory (McNicol, 2004) as a model. A signal detection model would posit that individuals learn over time how to calibrate their anxiety responses by learning which cues represent danger in the environment. The goal of development is to create an optimal strategy for doing so. Individuals in an ongoing process of exploration, assessment, and exploitation designed to assess the trade-off between obtaining a better estimate of the level of risk on one hand and maximizing current payoffs (the overall value of the threat assessment) on the other (Meacham & Bergstrom, 2016). This assessment of the resulting value of the trade-off solution is idiosyncratic to the individual and reflects the exigencies of their own life experiences. Meacham and Bergstrom suggest that as a result of this trade-off, certain people can become trapped in a state of self-perpetuating oversensitivity to threatening stimuli, even when individuals learn optimally. This phenomenon arises because when individuals become too cautious, they stop sampling the environment and fail to correct their misperceptions, whereas when individuals become too careless, they continue to sample the environment and soon discover their mistakes. Thus, oversensitivity to threats becomes common, whereas undersensitivity becomes rare. The more oversensitive you are to the threats, the greater your anxiety becomes. Again, remember that the perceived threats can be highly idiosyncratic, and a person may find something in the environment to be highly threatening when the vast majority of their peers would not reach the same conclusion. Thus, the attention system may be working as it should, but the resulting attention is miss/poorly allocated as a result of experience, reinforcement, or value calculations.

Attention is disrupted, but in a very different fashion from the disrupted attention caused by ADHD.

Emotional Dysregulation, Attention, and ADHD

As we saw in the chapter on "attention deficit hyperactivity disorder" (ADHD), dysregulated emotion is associated with this clinical condition. While it has long been accepted that many individuals with ADHD also have problems with emotion regulation, the nature of the relationship has not reached scientific consensus. Shaw et al. (2014) make several important points. First, the emphasis of emotional regulation as part of the DSM-based diagnostic manual has changed in relative importance to attention, with attention becoming the forefront issue. Second, they posit, we believe rightly so, that emotion regulation is a dimensional entity. In reviewing the association between emotion regulation and ADHD, they looked at the spectrum of emotion regulation within ADHD rather than as a comorbid condition. Using both quantitative and qualitative methods, three key findings/hypotheses emerge for consideration. First, emotion dysregulation is prevalent in ADHD throughout the lifespan and is a major contributor to overall impairment. Second, as we have just discussed, emotion dysregulation in ADHD may arise from deficits in orienting toward recognizing and/or allocating attention to emotional stimuli; these deficits implicate dysfunction within a striato-amygdalo-medial prefrontal cortical network. Third, while current treatments for ADHD often also amend emotion dysregulation, an emphasis on this combination of symptoms may generate additional clinical questions and could foster the development of novel therapeutic approaches.

Their research has also posited three models to explain the overlap between emotion dysregulation and ADHD: emotion dysregulation and ADHD are correlated but distinct *dimensions*, emotion dysregulation is a core diagnostic feature of ADHD, and the combination constitutes a nosologically distinct entity, which is distinct from both ADHD and emotion dysregulation alone. The differing predictions from each model can also guide research populations of clients with ADHD and emotion dysregulation. As it stands, once diagnosed with "attention deficit disorder," the clinical focus is on improving attention with the assumption that the emotional elements will be addressed almost by osmosis (Shaw et al., 2014).

Several psychological and neural processes underpin the overlap between ADHD and emotion dysregulation. Models describe both "bottom-up" processes that support or influence emotion regulation and "top-down" processes, such as the allocation of attention to emotionally arousing stimuli (Phillips et al., 2008).

As we have seen in the chapter on attentional control, deficits of attention are associated with many other clinical conditions including anxiety, depression, and schizophrenia. It is quite clear that emotional dysregulation is a dimensional trait that is not unique to ADHD; rather, it underpins the traditional divide between internalizing and externalizing diagnoses and, indeed, may partly explain their high correlation. For example, a study that contrasted 105 irritable, emotionally dysregulated children with ADHD and 395 nonirritable children with ADHD found higher rates not only of oppositional defiant disorder but also of depression and dysthymia in the group with irritability (Ambrosini et al., 2013).

Attention Types and Depression

A significant deficit of global attention has been reported in research detailing the attentional problems that characterize major depressive disorders (MDD). Specifically, MDD patients have shown significant deficits in psychomotor speed/ attention, auditory attention, visuospatial attention, and sustained visual attention with effect sizes ranging from small to moderate. Some studies have found deficits in sustained attention and some have not (Wang et al., 2020; Piani et al., 2022).

It is possible to characterize the specific characteristics of attention deficiencies in depression by focusing on several subdomains of experiences of individuals who experience depression. These subdomains include increased distractibility (selective attention impairment) to an inability to sustain focus (sustained attention impairment) or an inability to simultaneously monitor multiple channels of information (divided attention impairment (Keller et al., 2019)). It should be emphasized that depression is a highly heterogeneous disorder (Fried & Nesse, 2015) and different attention deficiencies, or no attention deficiencies, arising in the multiplicity of types of depression grouped together using the current nosology.

Selective attention is the ability to attend to important (task-relevant) information while ignoring distracting (task-irrelevant) information. Selective attention is also termed vigilance. Deficits in vigilance and psychomotor speed have been reported in individuals with major depressive disorder (Kertzman et al., 2010). Selective attention deficits are specifically associated with intrinsic hypo-connectivity of the frontoparietal attention network (Keller et al., 2020). They also found that selective attention was related to more severe negative bias (frequent misidentifications of neutral faces as sad and lingering attention on sad faces), relevant to clinical features of negative attributions and brooding. Finally, their research indicated that impairments of selective attention were independent of overall depression severity and of worrying or sleep problems.

Sustained attention refers to the ability to continuously attend to, or monitor, task-relevant information, usually assessed in the relative absence of distractions. In a meta-analysis (Piani et al., 2022), results showed the presence of alterations in the activation patterns of neural networks underpinning sustained attention in Major Depressive Disorder. The results were considered in line with research, suggesting that attentional deficiency was one of the key symptoms of depression. Data in this area is far from consistent with authors, suggesting that the homogeneity of the depression diagnosis occludes results.

Divided attention refers to multitasking and simultaneously attending to multiple sources of task-relevant information. Research has shown the importance of centrality of the prefrontal cortex in attending to multiple sensory modalities simultaneously (Godefroy & Rousseaux, 1996), conducting multiple tasks simultaneously (Richer et al., 1993) and switching attention between tasks (Owen et al., 1991). In the context of MDD, studies have pointed to impairments in performing two tasks simultaneously, and simultaneously attending both auditory and visual stimuli to detect targets (Keller et al., 2019).

There is research suggesting that depression is associated with poor attentional control (Yaroslavsky et al., 2019). Their research supports the idea that inflexible attention shifting with depressive affects is ascertained in naturalistic settings and corroborated the mediating role of rumination between attentional disengagement deficits and depression outcomes. The data pointed to the likelihood that rumination (not shifting attention from a negative stimulus) and a reduced capacity to disengage attention as appropriate to context (attention control deficits) have each been implicated in the etiology and maintenance of depressive disorders. Results also suggested that the relationship between attention inflexibility and rumination remains independent of depression severity and that these characteristics persisted after the depressive symptoms dissipated. In other words, it is quite possible that attentional predisposition, inflexibility, and rumination may be causal of a future depressive condition and not caused by the depression (Sanchez-Lopez et al., 2019). The data also suggested that these attentional characteristics remained after treatment and therefore should be targeted in their own right as part of a comprehensive treatment program for depression. They conclude by saying "our results imply that both constructs may continue to undermine emotional well-being even after depression remits and increase the risk for depression recurrence. Therefore, interventions that concomitantly target attention control deficits and rumination may hold the key to prolonging time spent in remission and to potentially reduce the risk for the recurrence of this episodic disorder" (p. 1067). These results are not isolated and have been replicated elsewhere (Hsu et al., 2015; Hu et al., 2017).

Emotional Dysregulation, Anxiety, and Attention

We have discussed the relationship of anxiety and attention control in the chapter on attention control and the reader should refer there for information regarding their interaction. This chapter references emotional dysregulation as defined by clinical condition.

Anxiety disorders are definitely highly comorbid with attention deficit hyperactivity disorders (ADHD). Together, these two conditions are the most prevalent psychiatric problems in childhood and adolescence, as demonstrated by recent epidemiological studies. These studies indicated that, these two disorders frequently occur in the same individual with the rate of comorbidity of 25% of cases in both epidemiological and clinical samples (Tannock, 2000.).

Dysregulated emotion in ADHD extends far beyond anxiety. In fact, three subtypes of ADHD: mild (normative emotional regulation), surgent (extreme levels of positive-approach motivation), and irritable (extreme levels of negative emotionality, anger, and poor soothability) have been proposed and supported in the research (Karalunas et al., 2014). The three subtypes were independent of established clinical groupings, showed stability over time, and were distinguished by unique patterns of cardiac physiological response, resting-state functional brain connectivity, and clinical outcomes. The current authors find this research very helpful in rounding out the multifaceted presentations of attention disorders. Clinicians dealing with persons qualifying as meeting the diagnostic criteria for an ADHD diagnosis are all likely familiar with parental/caretaker complaints of emotional dysregulation such as overly energetic emphasis on desired activities or pursuits, often described as incessant demandingness, or the irritability and behavioral dysregulation that can occur, for example, as a result of being disappointed or secondary to a schedule change. Regrouping emotionally can be difficult for these individuals. As ADHD is a heterogeneous diagnosis, for some, emotional regulation can fall within normal limits as well. However, for purposes of treatment, it is clearly necessary to best determine the type of presentation for the emotional dysregulation and the intensity in order to reduce the frequency and build resilience in the individuals. This would, of course, involve redirection of attention mechanisms depending upon where the dysregulation occurs in the cognitive process.

Generalized Anxiety Disorder

As the reader will likely be well informed at this point, it will come as no surprise that attention problems have been found in studies of generalized anxiety disorder. For the purposes of this discussion on emotional dysregulation, generalized anxiety disorder (GAD) can be thought of as a syndrome involving elevated intensity of subjective emotional experience, poor understanding of emotion, negative reactivity to emotional experience, and the use of maladaptive emotion management strategies, including an overreliance on cognitive control strategies such as worry (Turk et al., 2005). As noted earlier, this emotional arousal utilizes mental energy and hence we are not surprised by the impact on executive functions nor are we surprised by the attentional biases which occur.

Symptoms of emotion dysregulation in GAD, expressed across different psychophysiological axes, may reflect disruption of the degree to which the origin of symptoms overlap, and may inform the targeting of therapeutic interventions. Thayer and Lane (2000) posit a network model for the integration of autonomic, attentional, and affective control underlying emotion regulation and dysregulation They propose a model that integrates autonomic, attentional, and affective systems into a functional and structural network that is responsible for the regulation of emotion. The model emphasizes the relationship between attentional regulation and affective processes and describes underlying physiological systems that serve to integrate these functions in the service of self-regulation and adaptability of the individual. Their model is dynamic and involves feedback and feedforward circuits with specific focus on negative feedback mechanisms, inhibitory processes, and their role in response selection. These inhibitory processes are underpinned by negative feedback circuits that allow for the interruption of ongoing behavior and the redeployment of resources to other tasks (shifting). When these negative feedback mechanisms are compromised, positive feedback loops may develop as a result of the disinhibition. Thus, the relative sympathetic activation seen in anxiety disorders may represent disinhibition due to faulty inhibitory mechanisms. This model is, to the current authors, exemplifying the needs and benefits of looking at a cortical–subcortical model when attempting to better understand, and perhaps categorize, both diagnostic "labels" and intervention strategies.

Generalized anxiety disorder (GAD) is the most prevalent anxiety disorder. The most pervasive and behaviorally disruptive symptoms of GAD associated with significant impairment in daily functioning are excessive worry, difficulties concentrating, and feelings of physiological arousal (Makovac et al., 2020).

Individuals with GAD reported greater emotion intensity and fear of the experience of depression than persons with social anxiety disorder and non-anxious control participants. Of note, individuals with social anxiety disorder indicated being less expressive of positive emotions, paying less attention to their emotions, and having more difficulty describing their emotions than either persons with GAD or controls. For the purpose of this discussion, it is the issue of paying less attention to their emotions that concerns us. In general, it appears that most clinical disorders either hypersensitize or hyposensitize individuals to the experience of their emotions.

Attention and Bipolar Disorder

Research has described functional attentional abnormalities in adult bipolar disorder (BD), including both greater and reduced activity in lateral and dorsal PFC regions implicated in voluntary emotion regulation during voluntary attentional control paradigms and, largely, reduced activity in OFC and MdPFC implicated in automatic, emotion regulation during automatic attentional control paradigms (Blumberg et al., 2003; Kronhaus et al., 2006).

Using a rapid visual information processing (RVIP) continuous performance measure, Clark et al. (2002) found evidence of a relatively selective deficit on a task of sustained attention. This sustained attention deficit was potentially linked to vulnerability to bipolar disorder as it was present during periods of remission and a precursor trait to the development of the disorder. Their results also found that tests of verbal memory and attentional set shifting were sensitive to subtle shifts in mood in individuals with bipolar disorder. There have been suggestions that impaired set shifting be utilized as an objective marker of bipolar disorder, as opposed to depression, where it was associated with different comorbid factors (Maalouf et al., 2010).

Son et al. (2017) compared patterns of functional connectivity (FC) between affective and attention networks in adolescents with bipolar disorder (BD) and attention deficit hyperactivity disorder (ADHD). Their results showed that adolescents with mood fluctuation and attention problems showed increased functional correlation between two pairs of regions within the affective network (AFN): the left orbitofrontal cortex (OFC) to the left thalamus and the left OFC to the right thalamus. Further, adolescents with BD showed increased functional correlation between two pairs of regions compared to ADHD, specifically the right amygdala to the left temporoparietal junction (TPJ) and the right amygdala to the right TPJ. Adolescents with BD showed increased functional correlation within the attention network (ATN) as well as increased functional correlation between the ATN and the AFN, while those with ADHD showed decreased functional correlation within the ATN. The authors conclude that these disorder-specific features could be used as biomarkers for differentiating the two disorders. Here is direct neuroimaging evidence that there are deficits of attention, distinct from ADHD, that are present in mood dysregulation disorders and may characterize them as distinct features of specific disorders.

Of note, the aforementioned study was conducted on medication naïve participants. As it pertains to attentional processes in subtypes of bipolar disorder (BD), earlier studies suggested that BD type I may be related to worse attentional performance. Later research has disputed this fact and indicated both subtypes were associated with similar attentional impairment during euthymia, and that the differences between them may be related to antipsychotics, which were more likely to be used in BD type I (Pålsson et al. 2013).

Additional variables then come into play for individuals who are receiving psychotropic medication. For example, research with individuals with BD has also indicated that distinguishing drug-related deficits from disease-related deficits is difficult, and that drugs may have a certain degree of adverse effects on psychomotor speed and sustained attention (Batmaz et al., 2021).

Attentional bias is one of the areas that has been increasingly investigated in BD. In one of the earliest studies in this field, Jongen et al. (2007) found that mildly depressed clients focused their attention away from depression-related and positive words and reported this finding was mood state-dependent. They also found that the bias away from positive words could also be seen in euthymia, which may be part of a trait effect. For manic individuals, they were found to show no attentional vigilance to happy or sad expressions, but they avoided sad expressions and focused more on happy expressions in later emotion processing (Liu et al., 2018). This finding suggests distinct attentional biases may be present in both depressive and manic episodes.

In a review of neuropsychological assessments of the attentional processes of bipolar disorder (BD), target commission and omission errors were reported to be significantly present in the manic phase and target detection (omission) errors were reported as remaining in individuals with BD during remission (Clark & Goodwin, 2004). Sustained attention deficit in BD, which is persistent in euthymia and is exacerbated in the active disorder period, appears to be a state-modulated deficit (Najt et al., 2005). That is, the magnitude of presenting psychopathology-impacted attention.

Metacognitive theory (Wells, 2009) hypothesizes that psychopathology results from cognitive attentional syndrome (CAS). CAS is a set of psychological processes that can limit a person's thinking and attention. It is characterized by inflexible and recurrent thinking styles in response to negative thoughts, feelings, and beliefs, and can include worry, rumination, monitoring for threat, and unhelpful coping strategies such as avoidance, thought suppression, and substance use (Kowalski et al., 2019). The attention component of CAS refers specifically to attentional biases. Attentional biases reflect fixed attention on threats (threat monitoring) or focused attention on the self. Individuals with attentional biases become more aware of their internal experiences, many of which are essentially negative. The more focused they are on their thoughts, feelings, or sensations, the greater their reactions to such stimuli, creating a vicious cycle in which they begin to believe that their feelings reflect the reality and that feeling in that particular way is appropriate. Focusing on threat cues also increases the likelihood that they will feel distressed by them (Batmaz et al., 2021). As it pertains to the depressive phase of BD, metacognitive theory hypothesizes that the individual, in response to negative internal experiences that consist mostly of negative automatic thoughts or beliefs, focuses attention specifically to dwell on the past and/or on some real or imagined personal failure. This overthinking usually also involves focusing on worries about what may happen in the future, such as the recurrence of a new depressive episode. As a result, this negatively biased, selective, and self-referential attentional process extends the negative thinking style, eventually forming a cycle from which the individual cannot extricate themselves. Metacognitive theory also posits that the person holds some positive beliefs about the benefits of extended thinking, which may be associated with a belief that rumination will help to solve problems or to identify the root cause of negative feelings.

To address these issues therapeutically, the treating person using metacognitive theory works with the client to find ways to abandon any attentional biases, which generally consist of self-referential, self-focused attention, and threat monitoring, as well as ways to stop engaging with any dysfunctional, backfiring coping strategies (Batmaz et al., 2021). In metacognitive therapy (MCT), the treating professional focuses on identifying and reality-checking any positive and negative metacognitive beliefs so that any maintaining factor is targeted and eliminated (Wells & Fisher, 2015). Cognitive behavior therapy is particularly useful in this regard.

To address the requirements of individuals with BD, adjustments and modifications need to be considered. In a metacognitive case, conceptualization of a treating professional should consider focusing on positive trigger thoughts (or other positive internal experiences), as well as negative ones. In order to clarify the perseverative thinking style of a client, the treating professional may need to make note of any responses to positive affect and worries in addition to depressive rumination. In other words, the treating professional should encourage the individual to attend to various aspects of the triggers and their associated responses in order to help reorder and realign them.

The negative consequences of such thinking should also be addressed with consideration of any increase in self-reported energy levels, hypomanic/manic affective tendencies, hyperactivity, increased goal-directed activities, changes in sexual interest, impulsivity, increased self-esteem, and grandiosity, in addition to conventional thoughts, affect, and behaviors observed in depression. The treating professional and client must work collaboratively to identify ruminative thinking styles and worrying, with a special focus on positive rumination and how that may interfere with the client's goal to stay in remission. The role of positive rumination on unwanted consequences should be clarified, and postponement and/or abandonment of positive rumination must be reinforced, as it is for depressive rumination and worrying. The uncontrollability and dangerousness of all types of perseverative thinking styles need to be challenged. Modifying beliefs about the negative interpersonal consequences of perseverative thinking need to incorporate a discussion on the responses to positive affect in addition to the conventional discussion about depressive rumination and worrying. Any beliefs about the advantages of perseverative thinking need to be called into question, specifically questioning the perceived positive effects of positive rumination along with the conventional work on depressive rumination and worrying. Dysfunctional coping strategies involving thought control strategies, inactivity, self-harm, alcohol/substance use, emotional avoidance, social withdrawal, and so on may be targeted as they are with a client with depressive disorder. For the BD client, the treating professional also must acknowledge the possibility of increased goal-directed activity, hyperactivity, staying up all night to keep up with high expectations of self, impulsive behaviors, and related behaviors. Additionally, self-focused attention and constantly monitoring energy levels and mood states may be problematic for the client. Since a recurrence may be feared and warning signs for new episodes may be in focus, the client may wake up every day with an initial self/body scanning for any signs and symptoms suggestive of a new mood episode. The client needs to abandon such biased and backfiring threat monitoring strategies.

In other words, every aspect of the dysfunctional behavior should be elevated and made attentionally salient. The client is encouraged to attend to these aspects and their related corrective strategies throughout the course of treatment. Of course, the same general metacognitive strategy could be employed for other depressive disorders as well.

Emotional Dysregulation, Obsessive-Compulsive Disorder, and Attention

There are models based on an underlying hypothesis that nearly all psychiatric symptoms can be understood in terms of false inference. The premise of these models is that aberrant beliefs may reflect a failure to appropriately encode the confidence or precision that should be assigned to sensory evidence. Abramovitch et al. (2011) posit that the cognitive-behavioral model of obsessive-compulsive disorder has some empirical support, but does not fully explain the disorder. The cognitive behavioral model emphasizes the importance of dysfunctional beliefs in individuals. They provide the following categorical characteristics for consideration:

1. Obsessions about causing or failing to prevent harm and associated checking and reassurance seeking behavior.
2. Symmetry obsessions and associated ordering and counting rituals.
3. Contamination obsessions and associated washing and cleaning rituals.
4. Personally repugnant obsessions (concerning sex, religion, or violence).
5. Hoarding and collecting impulses.

Certainly, some of these characteristics may present for some of the individuals, but none are universal. We are again left with a very heterogenous diagnosis.

An alternative explanation posits that the primary pathology lies not in the way we explain the content of our perception, but in the way we balance competing prior beliefs and sensory input (Levy, 2018). Obsessive-compulsive disorder (OCD) is a good example. Levy would agree that the position that OCD involves giving catastrophic concerns excessive weight; however, he posits that the inappropriate assignment of probability actually reflects dysfunctionally heightened and directed attention.

Biological models of obsessive-compulsive disorder propose anomalies in the serotonin pathway and dysfunctional circuits in the orbito-striatal area and dorsolateral prefrontal cortex. Support for these models is mixed and they do not account for the symptomatic heterogeneity of the disorder.

While both models provide important facets, neither of the existing etiological models can fully explain the range of symptoms present in OCD, that is, the cognitive and symptom presentation components of the disorder. Levy holds that predictive coding addresses the collective presentation of the disorder.

Predictive Coding and OCD

Predictive coding is a theoretical framework that outlines how the brain generates internal models of the world and predicts sensory input based on prior beliefs (Gilbert et al., 2022). Utilizing a predictive coding account of OCD, it is possible to explain both the symptoms and the cognitive traits. The model is based in Bayesian probability theory and the Bayesian brain hypothesis (Knill & Pouget, 2004). The Bayesian brain hypothesis posits that perception is a constructive process that uses these models to encode sensory inputs and their causes. Basically, the predictive coding framework suggests that the brain functions to minimize surprise and uncertainty by actively generating explanations for encountered stimuli. This coding framework is based on the "Bayesian coding hypothesis": that the brain represents sensory information probabilistically, in the form of probability distributions.

Based on these concepts, OCD, at its core, involves heightened and dysfunctionally focused attention to normally unattended sensory and motor representations. In OCD, these representations have contents that predict catastrophic outcomes, and patients are disposed to engage in behaviors and mental rituals designed to forestall these outcomes or to produce more precise information. The same representations also cause the cognitive traits characteristic of individuals with OCD.

There is evidence that obsessive-compulsive disorder disrupts attention in another way. This research suggests that increasing working memory (WM) load reduces the executive control of attention. Hester and Garavan (2005) set up a series of three experiments to look at the relationship between working memory, or active processing, and measures of central executive tasks working under the premise that keeping the information in such a highly active state requires an individual to also control attention by attending to the task, blocking extraneous information, and switching as necessary. Their paradigm examined the relationship as measured via task-switching and inhibitory control paradigms (Hester & Garavan, 2005). In particular, their work indicated that the increased salience provided to the contents of WM through active rehearsal exerts a content-specific influence on attentional control. These findings when applied to the ruminations that characterize obsessive-compulsive disorder, wherein active rehearsal of evocative information (e.g., negative thoughts in depression or drug-related thoughts in addiction) in WM typically results from environmental cuing. When information maintained in WM is reencountered, it is harder for the person to exert executive control over it. That is, the ability to refer to task goals and inhibit the prepotent response became increasingly difficult. Due to the increased level of attention afforded to these cues, there is difficulty with suppressing the processing of these stimuli. This process hypothetically reinforces the maintenance of these cue induced ruminations items in WM. In this model, there is nothing wrong with attentional processing per se. The problem arises because attention is captured and directed to a suboptimal choice.

Predictive Coding Attention and the Etiology of OCD

Utilizing predictive coding, it is possible to generate an etiological model of OCD that has dysfunctional attentional processes at its core (Abramovitch et al., 2011). According to this model, OCD begins with elevated attention to action and thoughts. What precipitates this heightened attention likely changes from person to person. One of attention's basic functions is for the precise perception of inputs. Attention is also upregulated either by surprise or by increasing the stakes such that errors are costlier. Overestimation of threat can also be expected to upregulate attention, and there is a correlation between (relative) overestimation of threat and OCD. Anxiety might result from a heightened perception of responsibility or from an overestimation of threat (or from actual threats or trauma), and it also upregulates attention.

Once attention is upregulated, it becomes dysfunctionally focused. Since heightened attention is for increased precision, its failure to anxiety might itself constitute surprise/novelty. Surprise/novelty automatically engages attention, but since the problem arises from dysfunctionally enhanced attention, this essentially exacerbates the problem causing greater weight to be placed on signals that predict disaster.

In this model there is nothing wrong with the neural network of attention, rather it is the assignment of attention to suboptimal stimuli in the environment that is at issue. This suboptimal assignment of attention would cause the individual to appear inattentive in relation to the more adaptive optimal choice.

Predictive Coding and Other Disorders of Mental Health

Predictive coding has demonstrated efficacy in providing a framework for explaining psychopathologies including delusions (Hohwy, 2013; Harding et al., 2024) and depersonalization (Seth et al., 2011). A prevalent view is that many psychiatric conditions can be thought of as reflecting an imbalance between the precision afforded to sensory information and the precision of prior beliefs. Support for this position has been found in research on autism and schizophrenia (Adam et al., 2013).

OCD and Deficiencies in Shifting Attention

There is additional research that identifies difficulty in shifting attention as a core and perhaps distinguishing feature of OCD (Chamberlain et al., 2021; Deepthi et al., 2021). In a meta-analytic study of available research, they found attentional shift deficits to be a strong and reproducible finding (effect size medium–large) in OCD. Based on functional connectivity data, the OCD deficit was posited to reflect a dysfunction of fronto-striatal pathways. They concluded that impaired attentional shifting constituted a candidate vulnerability marker (or "endophenotype") for OCD. Similarly, cognitive inflexibility and deficient attentional shifting was found to be present in other disorders that represent obsessive-compulsive thinking, such as anorexia nervosa (Sternheim et al., 2022) and trichotillomania (Wilton et al., 2020).

Failure of Attention Allocation and Emotional Dysregulation as a Core Component of Emotional Disorders

There are models that hypothesize that poorly regulated attention underpins all mental disorders and that groupings of specific disorders are possible based on the type of attention disorder involved. One such model is provided by Sinnott-Armstrong et al. (2024). This model is a conceptual idea at this point but serves to make the point that a regrouping of disorders using a different diagnostic nosology is possible. The model distinguishes between two types of attention failure: a failure of tendencies and a failure of abilities. A failure of tendencies occurs when the individual repeatedly fails to direct or maintain their attention toward things that they have the most reason to pay attention to, or when they repeatedly fail to shift their attention from that they have reason not to pay attention to. Think about the child in math class who is looking at items on the teacher's desk rather than listening to the teacher explain how to solve the problem on the board. A failure of abilities occurs when the individual cannot sustain, direct, or shift their attention despite having valid reasons for doing so. This grouping aligns with the standard dysfunction-based model of the DSM-5 (Diagnostic and statistical manual of mental disorders (5th ed.), 2013). Obviously, their model of the attention underpinning of most mental disorders is based on top-down regulation. Based upon this distinction, it becomes possible to classify mental health disorders into two classes, the understanding of which may provide useful insight into new and effective treatments. It may also be possible to understand why some of these disorders demonstrate some degree of symptom overlap. They propose the following subdivisions:

Group 1. Failure to sustain attention to a stimulus in the environment when there is reason to pay attention to it.

Behavior	Emotion	Cognition
ADHD	Depression	Delusions

Group 2. Failure to shift attention when there is reason to do so.

Behavior	Emotion	Cognition
Eating Disorders	Phobias	Narcissism

For example, an individual with anorexia nervosa, in this model, cannot keep their thoughts from dwelling on body shape or weight and things that might affect it. This is because of the distorted perception of the world around them. The perception alone, however, is not sufficient. It is the inability to stop the negative thoughts that causes the disorder. Depression, on the other hand, results from the inability to focus on the more adaptive positive life circumstances that an individual encounters.

We would extend this point to include underlying common neural pathways that subserve these disorders and posit that it is possible to generate a newer nosology based on neural network characteristics as opposed to distinguishing behavioral characteristics. A first step in this direction was made with the introduction of the Research Domain Criteria (RDoC) model (Insel, 2014). The model provides views of mental health and psychopathology in the context of major domains of basic human neurobehavioral functioning. Rather than utilizing the traditional practice of symptoms for disease state identifications and then seeking specific pathologies, RDoC reverses the idea by considering psychopathology in terms of deviations from normal functions that could account for particular symptoms. These symptoms can occur on a transdiagnostic basis.

The RDoC model is based on the idea that both behavioral and biological aspects of functioning change and mature across the life span. Six major functional domains characterize various types of neurobehavioral functioning. The RDoC classifies mental disorders based on these six domains of human functioning and behavior:

1. Negative and positive valences
2. Cognitive systems (attention included)
3. Systems for social processes
4. Arousal/regulatory systems (attention included)
5. Sensorimotor systems
6. Neurobehavioral functioning

Different aspects of each domain are represented by psychological/biological dimensions, or constructs, which are considered along the full range of functioning from normal to abnormal. Equally important is the study of various aspects of the environment, including the physical environment, cultural components, and factors such as social determinants of health. Researchers are encouraged to measure and integrate multiple classes of data (units of analysis, e.g., behavior, genetics, physiology, and self-report data) to seek a comprehensive understanding of the construct(s) under study. As should be clear from the aforementioned, the RDoC model is currently for research purposes in order to better understand, classify, and treat psychopathologies. Research designs with such an integrative approach are, however, essential in order to obtain a comprehensive understanding both of normal functioning and of psychopathology. Humans are complex organisms. We have evolved to represent higher order complexities of cognitive functioning. It is also true that when the neurodevelopmental process goes awry, there is a limited number of ways in which the disturbances can manifest. Exploring and identifying the etiology, the combination of manifestations and dimensions of intensity is the sophisticated nosology that a sophisticated organism deserves and requires. While we do understand the demands of such a comprehensive system as currently aspirational, we can and must attempt to consider each of these elements and dimensions in order to better define and treat the dysfunctions associated with the issues presented to us as clinicians. We also believe that we owe it to our clients to provide them with an understanding of their areas of concern, which can be treated in a professional manner, comparable to other scientifically based healing professions.

It is possible to create new nosologies that will drive research with the potential to yield better understanding of these disorders and the potential to uncover new and more effective treatments.

References

Abramovitch, A., Dar, R., Hermesh, H., & Schweiger, A. (2011). Comparative neuropsychology of adult obsessive-compulsive disorder and attention deficit/hyperactivity disorder: Implications for a novel executive overload model of OCD. *Journal of Neuropsychology*. https://doi.org/10.1111/j.1748-6653.2011.02021

Adam, R., Stephan, K., Brown, H., Frith, C., & Friston, K. (2013). The computational anatomy of psychosis. *Frontiers in Psychiatry, 4*, 47. https://doi.org/10.3389/fpsyt.2013.00047

Ambrosini, P., Bennett, D., & Elia, J. (2013). Attention deficit hyperactivity disorder characteristics, II: Clinical correlates of irritable mood. *Journal of Affective Disorders, 145*, 70–76. https://doi.org/10.1016/j.jad.2012.07.014

Batmaz, S., Altinoz, S., & Sonkurt, H. (2021). Cognitive attentional syndrome and metacognitive beliefs as potential treatment targets for metacognitive therapy in bipolar disorder. *World Journal of Psychiatry, 11*(9), 589–604. https://doi.org/10.5498/wjp.v11.i9.589

Beard, C., Sawyer, A., & Hofmann, S. (2012). Efficacy of attention bias modification using threat and appetitive stimuli: A meta-analytic review. *Behaviour Therapy and Research, 43*(4), 724–740. https://doi.org/10.1016/j.beth.2012.01.002. Epub 2012 Jan 18.

Blumberg, H. P., Leung, H. C., Skudlarski, P., Lacadie, C. M., Fredericks, C. A., Harris, B. C., ... & Peterson, B. S. (2003). A functional magnetic resonance imaging study of bipolar disorder: state-and trait-related dysfunction in ventral prefrontal cortices. *Archives of General Psychiatry, 60*(6), 601–609.

Brazelton, T. B. (1973). *Neonatal behavioral assessment scale.* Spastics International Medical Publications: Heinemann Medical.

Brieant, A., King-Casas, B., & Kim-Spoon, J. (2022). Transactional relations between developmental trajectories of executive functioning and internalizing and externalizing symptomatology in adolescence. *Developmental Psychopathology, 34*(1), 213–224. https://doi.org/10.1017/S0954579420001054

Carleton, R. N., Sapach, M. J. T., Oriet, C., Duranceau, S., Lix, L. M., Thibodeau, M. A., ... & Asmundson, G. J. (2015). A randomized controlled trial of attention modification for social anxiety disorder. *Journal of Anxiety Disorders, 33*, 35–44.

Chamberlain, S., Solly, J., Hook, R., Vaghi, M., & Robbin, T. (2021). Cognitive inflexibility in OCD and related disorders. In N. Fineberg & T. Robbins (Eds.), *The neurobiology and treatment of OCD: Accelerating progress* (Current Topics in Behavioral Neurosciences) (Vol. 49). Springer, Switzerland AG. https://doi.org/10.1007/7854_2020_198

Clark, L., & Goodwin, G. M. (2004). State-and trait-related deficits in sustained attention in bipolar disorder. *European Archives of Psychiatry and Clinical Neuroscience, 254*, 61–68.

Clark, L., Iversen, S. D., & Goodwin, G. M. (2002). Sustained attention deficit in bipolar disorder. *The British Journal of Psychiatry, 180*(4), 313–319.

Derakshan, N., & Eysenck, M. (2009). Anxiety, processing efficiency, and cognitive performance: New developments from attentional control theory. *European Psychologist, 14*(2), 168–176. https://doi.org/10.1027/1016-9040.14.2.168

Deepthi, K., Roopesh, B. N., Balachander, S., Sagar, J. K. V., Kandavel, T., & Reddy, Y. J. (2021). Neuropsychological performance in youth with obsessive-compulsive disorder. *Journal of Psychiatric Research, 138*, 301–310.

Diagnostic and statistical manual of mental disorders. (2013). (5th ed.). American Psychiatric Association.

Dudeney, J., Sharpe, L., & Hunt, C. (2015). Attentional bias towards threatening stimuli in children with anxiety: A meta-analysis. *Clinical Psychology Review, 40*, 66–75. https://doi.org/10.1016/j.cpr.2015.05.007

Eysenck, M., & Derakshan, N. (2011). New perspectives in attentional control theory. *Personality and Individual Differences, 50*(7), 955–960. https://doi.org/10.1016/j.paid.2010.08.019

Fried, E., & Nesse, R. (2015). Depression is not a consistent syndrome: An investigation of unique symptom patterns in the STAR*D study. *Journal of Affective Disorders, 172*, 96–102. https://doi.org/10.1016/j.jad.2014.10.010

Gilbert, J., Wusinich, C., & Zarate, C. (2022). A predictive coding framework for understanding major depression. *Frontiers in Human Neuroscience, 16*. https://doi.org/10.3389/fnhum.2022.787495

Godefroy, O., & Rousseaux, M. (1996). Divided and focused attention in patients with lesion of the prefrontal cortex. *Brain and Cognition, 30*(2), 155–174. https://doi.org/10.1006/brcg.1996.0010

Gross, J. (2013). Emotion regulation: Taking stock and moving forward. *Emotion, 13*, 359–365. https://doi.org/10.1037/a0032135

Groves, N., Kofler, M., Wells, E., Day, T., & Chan, E. (2020). An examination of relations among working memory, ADHD symptoms, emotion regulation. *Journal of Abnormal Child Psychology, 48*, 525–537. https://doi.org/10.1007/s10802-019-00612-8

Harding, J., et al. (2024). A new predictive coding model for a more comprehensive account of delusions. *Personal View, 11*(4), 295–302. https://doi.org/10.1016/S2215-0366(23)00411-X

Hester, R., & Garavan, H. (2005). Working memory and executive function: The influence of content and load on the control of attention. *Memory and Cognition, 33*, 221–233. https://doi.org/10.3758/BF03195311

Hohwy, J. (2013). *The predictive mind.* Oxford University Press.

Hsu, K., Beard, C., Rifkin, L., Dillon, D., Pizzagalli, D., & Björgvinsson, T. (2015). Transdiagnostic mechanisms in depression and anxiety: The role of rumination and attentional control. *Journal of Affective Disorders, 188*, 22–27. https://doi.org/10.1016/j.jad.2015.08.008

Hu, B., Rao, J., Li, X., Cao, T. L., Majoe, D., & Gutknecht, J. (2017). Emotion regulating attentional control abnormalities in major depressive disorder: An event-related potential study. *Scientific Reports, 7*, 13530. https://doi.org/10.1038/s41598-017-13626-3

Insel, T. (2014). The NIMH research domain criteria (RDoC) project: Precision medicine for psychiatry. *The American Journal of Psychiatry, 171*(4), 395–397. https://doi.org/10.1176/appi.ajp.2014.14020138

Imperatori, C., Farina, B., Adenzato, M., Valenti, E. M., Murgia, C., Della Marca, G., ... & Ardito, R. B. (2019). Default mode network alterations in individuals with high-trait-anxiety: An EEG functional connectivity study. *Journal of Affective Disorders, 246*, 611–618.

Jongen, E. M., Smulders, F. T., Ranson, S. M., Arts, B. M., & Krabbendam, L. (2007). Attentional bias and general orienting processes in bipolar disorder. *Journal of Behavior Therapy and Experimental Psychiatry, 38*(2), 168–183.

Karalunas, S., Fair, D., Musser, E., Aykes, K., Iyer, S., & Nigg, J. (2014). Subtyping attention-deficit/hyperactivity disorder using temperament dimensions: Toward biologically based nosologic criteria. *JAMA Psychiatry, 71*, 1015–1024. https://doi.org/10.1001/jamapsychiatry.2014.763

Keller, A., et al. (2019). Paying attention to attention in depression. *Translational Psychiatry, 9*, 279. https://doi.org/10.1038/s41398-019-0616-1

Keller, A. S., Ball, T. M., & Williams, L. M. (2020). Deep phenotyping of attention impairments and the 'Inattention Biotype'in Major Depressive Disorder. *Psychological Medicine, 50*(13), 2203–2212.

Kertzman, S., Reznik, I., Hornik-Lurie, T., Weizman, A., Kotler, M., & Amital, D. (2010). Stroop performance in major depression: Selective attention impairment or psychomotor slowness? *Journal of Affective Disorders, 122*(1–2), 167–173. https://doi.org/10.1016/j.jad.2009.08.009

Knill, D., & Pouget, A. (2004). The Bayesian brain: The role of uncertainty in neural coding and computation. *Trends in Neuroscience, 27*, 712–719. https://doi.org/10.1016/j.tins.2004.10.007

Kowalski, J., Wypych, M., Marchewka, A., & Dragan, M. (2019). Neural correlates of cognitive-attentional syndrome: An fMRI study on repetitive negative thinking induction and resting state functional connectivity. *Frontiers in Psychology, 10*. https://doi.org/10.3389/fpsyg.2019.00648

Kronhaus, D. M., Lawrence, N. S., Williams, A. M., Frangou, S., Brammer, M. J., Williams, S. C., ... & Phillips, M. L. (2006). Stroop performance in bipolar disorder: further evidence for abnormalities in the ventral prefrontal cortex. *Bipolar Disorders, 8*(1), 28–39.

Kuckertz, J. M., Amir, N., Boffa, J. W., Warren, C. K., Rindt, S. E., Norman, S., ... & McLay, R. (2014). The effectiveness of an attention bias modification program as an adjunctive treatment for post-traumatic stress disorder. *Behaviour Research and Therapy, 63*, 25–35.

Levy, N. (2018). Obsessive–compulsive disorder as a disorder of attention. *Mind and Language, 33*(1), 3–16. https://doi.org/10.1111/mila.12172

Liu, J. C., Liu, K., & Zhao, J. (2018). Event detection via gated multilingual attention mechanism. In *Thirty-second AAAI conference on artificial intelligence*. Association for the Advancement of Artificial Intelligence.

Maalouf, F. T., Klein, C., Clark, L., Sahakian, B. J., LaBarbara, E. J., Versace, A., ... & Phillips, M. L. (2010). Impaired sustained attention and executive dysfunction: bipolar disorder versus depression-specific markers of affective disorders. *Neuropsychologia, 48*(6), 1862–1868.

Makovac, E., Fagioli, S., Rae, C. L., Critchley, H., & Ottaviani, C. (2020). Can't get it off my brain: Meta-analysis of neuroimaging studies on perseverative cognition. *Psychiatry Research: Neuroimaging, 295*, 111020. https://doi.org/10.1016/j.pscychresns.2019.111020

Marquis, W. A., Noroña, A. N., & Baker, B. L. (2017). Developmental delay and emotion dysregulation: Predicting parent–child conflict across early to middle childhood. *Journal of Family Psychology, 31*(3), 327–335. https://doi.org/10.1037/fam0000267

McNicol, D. (2004). *A primer of signal detection theory*. Psychology Press. https://doi.org/10.4324/9781410611949

Meacham, F., & Bergstrom, C. (2016). Adaptive behavior can produce maladaptive anxiety due to individual differences in experience. *Evolution, Medicine, and Public Health, 2016*(1), 270–285. https://doi.org/10.1093/emph/eow02

Najt, P., Glahn, D., Bearden, C., Hatch, J. P., et al. (2005). Attention deficits in bipolar disorder: A comparison based on the Continuous Performance Test. *Neuroscience Letter, 379*, 122–126. https://doi.org/10.1016/j.neulet.2004.12.051

Owen, A. M., Roberts, A. C., Polkey, C. E., Sahakian, B. J., & Robbins, T. W. (1991). Extra-dimensional versus intradimensional set shifting performance following frontal lobe excisions, temporal lobe excisions or amygdalohippocampectomy in man. *Neuropsychologia, 29*, 993–1006.

Pålsson, E., Figueras, C., Johansson, A. G., Ekman, C. J., Hultman, B., Östlind, J., & Landén, M. (2013). Neurocognitive function in bipolar disorder: a comparison between bipolar I and II disorder and matched controls. *BMC Psychiatry, 13*, 1–9.

Paulus, F., Ohmann, S., Plenar, P., & Popow, C. (2021). Emotional dysregulation in children and adolescents with psychiatric disorders. A narrative review. *Frontiers in Psychiatry, 12*. https://doi.org/10.3389/fpsyt.2021.628252

Pergamin-Hight, L., Naim, R., Bakermans-Kranenburg, M. J., van IJzendoorn, M. H., & Bar-Haim, Y. (2015). Content specificity of attention bias to threat in anxiety disorders: A meta-analysis. *Clinical Psychology Review, 35*, 10–18.

Phillips, M., Ladouceur, C., & Drevets, W. (2008). A neural model of voluntary and automatic emotion regulation: Implications for understanding the pathophysiology and neurodevelopment of bipolar disorder. *Molecular Psychiatry, 13*, 829, 833–857.

Piani, M., Maggioni, E., Delvecchio, G., & Brambilla, P. (2022). Sustained attention alterations in major depressive disorder: A review of fMRI studies employing Go/no-Go and CPT tasks. *Journal of Affective Disorders, 303*, 98–113. https://doi.org/10.1016/j.jad.2022.02.003

Richer, F., Decary, A., Lapierre, M., Rouleau, I., Bouvier, G., & Sainthilaire, J. (1993). Target detection deficits in frontal lobectomy. *Brain and Cognition, 21*(2), 203–211. https://doi.org/10.1006/brcg.1993.1016

Roberts, C. S. (2023). Impact and centrality of attention dysregulation on cognition, anxiety and low mood in adolescents. *Scientific Reports, 13*.

Romer, A., & Pizzagalli, D. (2021). Is executive dysfunction a risk marker or consequence of psychopathology? A test of executive function as a prospective predictor and outcome of general psychopathology in the adolescent brain cognitive development study. *Developmental Cognitive Neuroscience., 51*, 100994. https://doi.org/10.1016/j.dcn.2021.100994

Ross, A., & Thompson, R. (2011). Emotion and emotion regulation: Two sides of the developing coin. *Emotion Review, 3*, 53–61. https://doi.org/10.1177/1754073910380969

Sanchez-Lopez, A., Koster, E., & Van Put, J. (2019). Attentional disengagement from emotional information predicts future depression via changes in ruminative brooding: A five-month longitudinal eye-tracking study. *Behavior Research and Therapy, 118*, 30–42. https://doi.org/10.1016/j.brat.2019.03.013. Epub 2019 Mar 27.

Seth, A., Suzuki, K., & Critchley, H. (2011). An interoceptive predictive coding model of conscious presence. *Frontiers in Psychology, 2*, 395. https://doi.org/10.3389/fpsyg.2011.00395

Shaw, P., Stringaris, A., Nigg, J., & Leibenluft, E. (2014). Emotion dysregulation in attention deficit hyperactivity disorder. *The American Journal of Psychiatry, 171*, 276. https://doi.org/10.1176/appi.ajp.2013.13070966

Shi, R., Sharpe, L., & Abbott, M. (2019). A meta-analysis of the relationship between anxiety and attentional control. *Clinical Psychology Review, 72*, 101754.

Sinnott-Armstrong, W., Soter, L. K., & Summers, J. S. (2024). Mental Disorders as Failures of Attention. Crítica. Revista Hispanoamericana de Filosofía, 17–44.

Solmi, M., Dragioti, E., Arango, C., Radua, J., Ostinelli, E., Kilic, O., ... & Fusar-Poli, P. (2021). Risk and protective factors for mental disorders with onset in childhood/adolescence: an umbrella review of published meta-analyses of observational longitudinal studies. *Neuroscience & Biobehavioral Reviews, 120*, 565–573.

Son, Y. D., Han, D. H., Kim, S. M., Min, K. J., & Renshaw, P. F. (2017). A functional connectivity comparison between attention deficit hyperactivity disorder and bipolar disorder in medication-naïve adolescents with mood fluctuation and attention problems. *Psychiatry Research: Neuroimaging, 263*, 1–7.

Sternheim, L., van Passal, B., Dingermans, A., Cath, D., & Danner, U. (2022). Cognitive and experienced flexibility in patients with anorexia nervosa and obsessive compulsive disorder. *Frontiers Psychiatry, 16*. https://doi.org/10.3389/fpsyt.2022.868921

Stringaris, A., et al. (2010). What's in a disruptive disorder? Temperamental antecedents of oppositional defiant disorder: Findings from the Avon longitudinal study. *Journal of the American Academy of Child & Adolescent Psychiatry, 49*(5), 474–483.

Sylvester, C., Corbetta, M., Raichle, M., Rodebaugh, T., Schlaggar, B., Sheline, Y., et al. (2012). Functional network dysfunction in anxiety and anxiety disorders. *Trends in Neuroscience, 35*(9), 527–535. https://doi.org/10.1016/j.tins.2012.04.012

Tannock, R. (2000). Attention-deficit/hyperactivity disorder with anxiety disorders. In I. T. Brown (Ed.), *Attention-deficit disorders and comorbidities in children, adolescents, and adults* (pp. 125–170). American Psychiatric Publishing, Inc.

Thayer, J. F., & Lane, R. D. (2000). A model of neurovisceral integration in emotion regulation and dysregulation. *Journal of Affective Disorders, 61*(3), 201–216.

Thompson, R. (1994). Emotion regulation: A theme in search of definition. *Monographs of the Society for Research in Child Development, 59*(2/3), 25–52.

Turk, C. L., Heimberg, R. G., Luterek, J. A., Mennin, D. S., & Fresco, D. M. (2005). Emotion dysregulation in generalized anxiety disorder: A comparison with social anxiety disorder. *Cognitive Therapy and Research, 29*, 89–106.

Wang, X., Zhou, W., & Zhu, X. (2020). Attention deficits in adults with Major depressive disorder: A systematic review and meta-analysis. *Asian Journal of Psychiatry, 53*, 102359. https://doi.org/10.1016/j.ajp.2020.102359

Wassserman, L. W. (2024, April 19). *Motor network and implications for neurodevelopmental disorders*.

Wehmeier, P. M., Schacht, A., & Barkley, R. A. (2010). Social and emotional impairment in children and adolescents with ADHD and the impact on quality of life. *Journal of Adolescent Health, 46*(3), 209–217.

Wells, A. (2009). *Metacognitive therapy for anxiety and depression*. Guilford Press.

Wells, A., & Fisher, P. (2015). *Treating depression: MCT, CBT, and third wave therapies*. ohn Wiley & Sons.

Wilton, E. P., Flessner, C. A., Brennan, E., Murphy, Y., Walther, M., Garcia, A., ... & Freeman, J. B. (2020). A neurocognitive comparison of pediatric obsessive-compulsive disorder and trichotillomania (hair pulling disorder). *Journal of Abnormal Child Psychology, 48*, 733–744.

Yaroslavsky, I., Allard, E., & Snachez-Lopez, A. (2019). Can't look Away: Attention control deficits predict Rumination, depression symptoms and depressive affect in daily Life. *Journal of Affective Disorders, 245*, 1061–1069. https://doi.org/10.1016/j.jad.2018.11.036

Zorcec, T., & Jordanova, N. (2010). ADHD as an executive dysfunction. *Contributions, Section Biological, Medical Science, XXXI*(2), 171–181. 159.922.7:616.8-008.6.

Chapter 12
Toward a Spectrum of Attention Disorders

When we conceptualized the writing of this book, we identified several main ideas or hypotheses that we thought were important to discuss. The first idea was the notion that from our developmental perspective, attention is a highly foundational process that is present in the neonate and progresses, along with the other central executive skills and motor skills, in order to integrate into a highly sophisticated organism. From this perspective, it becomes clear that attention is inherent in everything humans do. Following this line of reasoning, attention is involved in both neurotypical and neurodivergent development.

This then brings us to what happens when this particular process we call attention is compromised. We posit that such a fundamental construct cannot functionally exist discretely. As a necessary goal of development, these foundational processes continually integrate themselves with other processes. As a result, when impacted, it can present within and across many disorders. Therefore, all disorders of attention do not represent aspects of attention deficit hyperactivity disorder (ADHD). That is, there are problems of attention, associated with other clinical conditions, that arise independently of attention deficit hyperactivity disorder. We believed that these other attention issues were poorly articulated, and perhaps often mislabeled or overlooked.

The second idea was that the classification of ADHD, as far as the DSM-5 defines it, was overly broad and too heterogeneous to be of scientific value going forward. A related idea was that grouping all types of attentional deficiencies under the rubric of ADHD resulted in dilution of certain characteristics and research findings, making the utility of findings based on heterogeneous samples less clinically utilitarian.

Finally, related to this last idea, we wished to put forward the idea that attention deficit hyperactivity disorder-inattentive type was probably a separate disorder with unique neural network underpinnings.

While researching and writing this book, some additional ideas presented themselves and provided the opportunity to think about things quite differently. Foremost

T. Wasserman, L. D. Wasserman, *Disorders of Attention*, Neural Network Model: Applications and Implications, https://doi.org/10.1007/978-3-031-78732-4_12

among these was the idea that attentional functioning (and poor functioning) could provide a meaningful, and empirically valid, way to understand the difference between traditional disorders of mental health and pave the way for a new nosology. That there were commonalities of poor attentional allocation that underpinned certain disorders of mental health and differentiated them from other groups of mental health disorders. Finally, there is the idea that in many instances, attention per se is not disordered at all, but rather it is misapplied or less than maximally efficiently utilized.

We have addressed these ideas in the book and will take each of the above ideas and review them here. We hope that, in doing so, we can provide a fresh perspective on attention and pave the way forward to the development of a nosology based on neural network functioning as opposed to only using behavioral characteristics.

All deficits of attention do not represent aspects of attention deficit hyperactivity disorder.

As we have demonstrated in the chapter on emotional dysregulation and attention, there are clinical disorders that have associated attentional processing disruption. The co-occurrence of attention dysregulation is associated with a wide range of adverse outcomes related to emotional dysregulation, psychopathological features, and executive functioning (EF) impairments (Roberts et al., 2023). The disrupted attention in these disorders is underpinned by different subroutines of the overall attention network system. These mood disorder specific disruptions of attention frequently do not meet the criteria for a diagnosis of attention deficit disorder although in clinical practice they might be diagnosed as comorbid ADHD. Disorders with at least some portion of affected individuals having concomitant attention processing issues include major depressive disorder (Wang et al., 2020), bipolar disorder (Clark & Goodwin, 2004), general anxiety disorder and obsessive-compulsive disorder (Abramowitz, 2009), among others.

We have pointed out that often, these associated attentional problems are relegated, and to some extent minimized, when treatment is formulated to target what is considered the primary disorder, most pronounced symptom, or most disruptive symptom. Treatment usually targets the presumed primary diagnosis with the assumption that the associated attention problem will resolve if the primary symptomatology is addressed. That is often not the case, and some data suggests that treatment should focus on these attention issues in their own right. We have suggested that new targeted treatments for the issue might be advantageous although many of these attention problems do respond positively to extant treatments for ADHD. There is research support for this position (Keller et al., 2019). When discussing treatment for depression, they concluded that in many instances, individuals with major depressive disorder (MDD) complain of concentration difficulties that adversely impact their day-to-day functioning, and that these attention problems, specifically associated with depression, are not alleviated by current first-line treatments for depression. This occurs because, despite attention's influence on multiple features of cognitive and emotional functioning, and the actual inclusion of concentration difficulties in the diagnostic criteria for MDD, the focus of depression as a disorder is typically on the mood aspects, with attention features considered less of an imperative for research investigation or targeted treatment. They suggested that there are potential pharmacological, brain stimulation, and behavioral interventions and their effects on subdomains of goal-directed attention (selective, sustained, and divided) specifically target the attentional difficulties associated with depression.

Among the behavioral interventions, mindfulness meditation involves rehearsing the skill of selectively attending to one sensation (e.g., breath) while ignoring distracting thoughts. It has been shown to improve selective attention measured independently in addition to improving mood-related symptoms of MDD.

Another potential treatment for attentional problems is computerized cognitive training apps. Generally, these utilize cognitive exercises or immersive video games. These apps have been utilized as treatments for attentional and mood-related symptoms in MDD (Anguera et al., 2017). Both MM and computer training applications have potential uses with the attentional problems associated with other clinical disorders as well.

All of this suggests that the practicing clinician must be cognizant of the types of attentional processing issues that impact their clients. That they design their treatment protocols for common clinical conditions recognizing that they also need to address specifically the associated attentional issues. For .example, if a clinician is treating a person with obsessive-compulsive disorder (OCD), they need to consider the predictive coding distortions that are part of the cognitive presentation of this disorder: the brain actively predicts sensory input instead of passively registering it, so that in OCD processing is the idea that OCD begins with elevated attention to specific actions and thoughts. The clinician should ascertain which stimuli precipitate this heightened attention, and understand that the selection of precipitating stimuli is idiosyncratic from person to person. The practicing clinician should be cognizant that ideally, attention's basic function is for the precise perception of inputs. Misperception or poor predictive coding can result in hyperfocus on specific, and in the case of OCD, suboptimal features of the environment. Attention is also upregulated, with increased intensity, either by surprise or by increasing the stakes. In the case of OCD, errors can be costlier in a multitude of ways ranging from exaggerated emotional responses to missing important environmental stimuli. Overestimation of threat can also be expected to upregulate attention, and there is a correlation between (relative) overestimation of threat and OCD. Anxiety might result from a heightened perception of responsibility or from an overestimation of threat (or from actual threats or trauma and anxiety also upregulates attention. All of this leads to dysregulated attention or, may in fact be caused by dysregulated or poorly allocated attention. There are two immediate implications. The first is that none of the attentional processing issues we are discussing here represent attentional dysregulation caused by whatever ADHD is. Yet, our current system of diagnosis might encourage the practicing clinician to diagnose it as a comorbid condition rather than a core component of the OCD itself. The second is that by concentrating on the catastrophic cognitions or compulsive rituals, we are not alerting the patient to the attentional patterns they have been, and are engaging in, which perpetuates the disorder.

Defining Attention Somewhat Differently: A Foundational Element

Perhaps much of the confusion surrounding attention comes from the fact that we are defining it in its mature form. We look at attention and try to study attention as a discrete domain. However, as the current authors have pointed out in multiple presentations and publications, attention begins not as a primitive reflex, but as a foundational reflex we call alerting. It is purposeful and present from birth as noted by the neonate's attempts to alert to and locate caretaking and nourishment. In order to survive, the neonate must attend to environmental sources of caretaking. Attention then is already a binding, foundational element of sensory-motor prediction and reward. The initializing of this foundational reflex incorporates all domains and factors discussed in the adult literature including salience, prediction, motivation, reward, and emotional regulation. In short, it is a foundational element of all development, including neurodevelopment, yet to come. It is therefore a core component of sound neurodevelopment and a core component of pathologies when neurodevelopment goes awry.

Toward this point, some recent neuroimaging research bears mentioning. There has been interest in ascertaining the network dysfunction associated with behavioral addictions (BA), non-substance abuse behavior, but rather behavior such as excessive gambling, excessive shopping, exercising, and so on. While the preponderance of such research is done while the subject is in a resting state, Chen et al. (2024) looked to better assess functional coordination among large-scale networks. Individuals within the BA category have been shown to exhibit brain activation within the fronto-pariatal network (FN), the dorsal attention network (DAN), the affective network (AN), the ventral attention network (VAN), the default mode network (DMN), and the somatosory network (SS). As a reminder, FN, also called the central executive network, is involved with the top-down regulation of emotion and attention. The results demonstrate hyperconnectivity within the FN, hyper- and hypoconnectivity within the VAN, hyperconnectivity between the VAN and FN, and hyperconnectivity between the DMN and FN. The implications are clear in that neuro-involvement is across systems, and the current authors would emphasize that because the trajectory for neurodevelopment is integration for efficiency, the impact will be across domains.

Asking a Different Question

Asking whether the attentional disruption in a particular client represents attention deficit disorder results in a binary choice. It does or does not. The better question to ask is what kind of attentional disruption is occurring. There are attentional aspects to most, if not all, disorders of emotional regulation. These attention problems are integral to the disruption of behavior found in the various disorders. The attention problems in various categories of disorders differ in their underlying neural network composition. As we have seen, attention can be dysfunctional along a number of dimensions not typically considered in a typical clinical setting. Attention might be disrupted by a failure of the bottom-up, visually based orienting system. It might be disrupted by numerous top-down variables including salience issues, value-laden reinforcement calculations, misdirected attention based on history of reward, or reencountering items currently held in working memory. In fact, in some disorders, attention may not be disrupted at all, it may be misdirected, misallocated, or misapplied. In recognizing the breath and scope of attentional dysregulation, we acknowledge that there are many more disorders of attention functioning and/or usage that require clinical research into the development of more effective interventions. This of course leads to the question concerning the nature of ADHD. If all of the various problems of attention are not ADHD, what do we call them? How do we reliably assess them? Finally, if all of these issues are not ADHD, what exactly is ADHD?

The classification of ADHD, as far as the DSM-5 defines it, is overly broad and too heterogeneous to be of scientific value going forward.

There is little doubt that the idea of an entity called attention deficit hyperactivity disorder is clinically relevant. It has spawned a multiplicity of treatments that have resulted in clinically beneficial outcomes for millions of people. Nevertheless, problems abound. First and foremost, its behavioral criterion and the system that contains it results in a cluster of extremely heterogeneous issues being condensed into one diagnostic entity (Luo et al., 2019; Dias et al., 2015). Their research indicated that ADHD has multifactorial causation, leading to the heterogeneity of the disorder. The multifactorial causation of ADHD causes significant problems for research, namely issues for reliability and validity (Salum et al., 2018). Their research found 116,200 possible combinations of symptoms to achieve DSM symptomatic threshold for categorical ADHD diagnosis. Of that number, 173 combinations were found in two independent samples, and only four were replicated in both samples (2.3%). They indicated that the number of ADHD symptoms was a poor indicator of variation in the general ADHD latent trait. Their research indicated that there is a wide variation in severity in subjects, showing the same symptom count. Finally, as we have stressed earlier, ADHD symptoms did not have specific profiles of associations with any of the neurocognitive processes. This is clearly a problem as, even taking the lower number 173, it is very unlikely that there are 173 different types of ADHD. As pertains to reliability, the authors concluded that results from a reliability analysis demonstrated that increasing the number of symptoms increased reliability of the latent trait. They posited that this finding was consistent with the concept that part of variability is due to a measurement error, instead of true heterogeneity. However, it is possible that general effects are being found in the literature simply because they are more reliable and not because specific effects are not real. They pointed out that lack of reliability attenuates effect sizes and decreases the power of statistical tests, both of which compromise the ability to provide the evidence necessary to validate specific contributions.

This heterogeneity is also evidenced by the diversity of ADHD's "psychiatric comorbidities. A wide range of comorbid behavioral and psychiatric conditions are associated with ADHD, including learning disabilities, language disorders, mood disorders, anxiety, and conduct/oppositional disorder" (Reale et al., 2017). These comorbid problems confound research and, as we have seen, can complicate both diagnosis and treatment of ADHD.

In addition, varied clinical profiles, patterns of neurocognitive impairment and developmental trajectories, and the wide range of structural and functional brain anomalies occur within the existing ADHD structure. The etiological heterogeneity in biological and environmental factors is reflected in variation in neural correlates (Costa Dias et al., 2013; Nigg & Casey, 2005) and results in the numerous and diverse cognitive and behavioral profiles and developmental trajectories of the disorder.

While the DSM-5 behaviorally based diagnosis of ADHD provides the benefit of a common language and standard criteria for identification of some presentations of ADHD and their subtypes, there have been elements of and potential subtypes that have been ignored. For example, emotion lability-based subtypes and reward recognition issues have been suggested by research. Deficient emotional self-regulation has been suggested to be a core component of ADHD (Barkley & Fischer, 2010). For example, (Graziano & Garcia, 2016) in summarizing the results of their meta-analysis of 77 studies with a total of 32,044 participants reported the associations of ADHD with impaired emotional reactivity/negativity/liability, and empathy/callous–unemotional traits of emotional regulation. Other research using the "research design criteria" has generated data that has suggested different subgroupings. They identified subgroups based on the data from the Child Behavior Checklist (CBCL) (Achenbach, 1991) and the Behavior Rating Inventory of Executive Function (BRIEF) (Gioia et al., 2002). The subgroups' clinical characteristics and functional impairments were assessed using the WEISS Functional Impairment Rating Scale-Parent Report (WFIRS-P) (Weiss, 2000) and the Conners Parent Symptom Questionnaire (PSQ) (Connors, 2022). Their cluster analysis yielded four subgroups: (1) ADHD with severe impairment in psychopathology and executive functions (EF), (2) ADHD with mild executive dysfunctions and normal-level psychopathology, (3) ADHD with severe externalizing problems, and (4) ADHD with severe executive dysfunctions. "These subgroups showed different clinical characteristics and degrees of functional impairment. The EF impairment group displayed more serious learning problems and worse life skills than the externalizing group. The two groups with externalizing problems (i.e. the severe impairment group and the externalizing group) both exhibited higher rates of the combined subtype of ADHD and higher rates of comorbid ODD. Different subtypes of ADHD displayed different profiles of internalizing and externalizing problems and levels of executive dysfunctions. In particular, the subtype with severe impairment in EF exhibited more learning problems and worse life skills, suggesting EF is a critical target for intervention in children with ADHD."

In addition to the behavioral heterogeneity, there is significant genetic heterogeneity in the cluster of disorders that fall under the ADHD umbrella (Coghill & Banaschewski, 2009; LaBianca & Brikell et al., 2024). Molecular genetic studies (Thapar et al., 2006) have pointed to a range of potential candidate genes. From these studies, we have learned that the genetics of ADHD and its subtypes are characterized by a number of subtype specific genes, each of which makes a small but significant contribution to the overall risk of the disorder becoming behaviorally manifest. Several regions have been supported across several studies (bin 16.4, 5p13, 11q22–25, 17p11). The contribution of several candidate genes (*DRD4*, *DRD5*, *DAT1*, *HTR1B*, and *SNAP25*) has been supported by meta-analysis (Li et al., 2006). Recent research has expanded on this group of candidate genes. They utilized a genome-wide association study meta-analysis of ADHD comprising 38,691 subjects with ADHD and 186,843 controls. The results indicated 27 genome-wide significant loci, highlighting 76 potential risk genes enriched among genes expressed particularly in early brain development. The genetic risk of ADHD was associated with several brain-specific neuronal subtypes and midbrain dopaminergic neurons. In exome-sequencing data (The exome is the part of the genome codes for proteins. The exome makes up about 1% of the genome but contains around 85% of known disease-related variants) from 17,896 individuals, they identified an increased load of rare protein-truncating variants in ADHD for a set of risk genes enriched with probable causal common variants, potentially implicating *SORCS3* in ADHD by both common and rare variants. 84–98% of ADHD-influencing variants were shared with other psychiatric disorders. In addition, common-variant ADHD risk was associated with impaired complex cognition such as verbal reasoning and a range of executive functions, including attention.

Even among the existing behaviorally defined DSM-5 categories, there is considerable genetic difference. For example, Kunsti et al. (2014), based upon their review of research, concluded that "both shared and unique genetic risk factors underlie the two symptom domains of attention deficit hyperactivity disorder (ADHD): inattention and hyperactivity-impulsivity. The developmental course and relationship to co-occurring disorders differs across the two symptom domains, highlighting the importance of their partially distinct etiologies" (p. 127).

As we have discussed, three different types of developmental trajectories have been identified within the ADHD population. These trajectories differ both in network properties and genetic contribution. Their research has identified these three distinct trajectories as ascending-high, stable-low, and descending-medium. These trajectories differ from childhood through adolescence. These classifications delineate the patterns by their ADHD status as remission, persistence, or late onset. Specific attention hub regions provided the neural underpinnings for these three developmental trajectories. These regions encompassed the left anterior cingulate cortex/medial prefrontal cortex (ACC/mPFC), responsible for inhibitory control; the right inferior parietal lobule (IPL), which facilitated conscious focus on exogenous stimuli, and the bilateral middle frontal gyrus/precentral gyrus (MFG/PCG) responsible for regulating both dorsal and ventral attention networks. The MFG/PCG also played a central role in flexible modulation of endogenous and extrinsic attention. As to genetic contribution, individuals in the ascending-high group exhibited the highest polygenic risk score (PRS) for ADHD, followed by those in the descending-medium group, with individuals in the stable-low group displaying the lowest PRS. Both ascending-high and descending-medium groups had significantly higher PRS compared to the stable-low group. Polygenic risk scores are numeric estimates of an individual's genetic likelihood of developing a disease or trait at birth.

The Heterogeneity Problem

We have pointed out that the heterogeneity of the class of disorders that are considered a problem for ADHD potentially confounds research. We believe that the imprecise nature of the current, behaviorally grounded (phenotypical) psychiatric nosology restricts progress toward characterizing/treating mental health disorders. One issue is the "heterogeneity problem" (Feczko et al., 2019). That is, different causal mechanisms may relate to the same disorder, and multiple outcomes of interest can occur within one individual. In addition, lumping disorders together likely obscures potentially significant findings for relevant subgroupings. The core concerns with the DSM and ICD nosology for psychiatry are well known and include both over- and underspecificity, that is: (1) heavy overlap among the "disorders" and shared biological features, indicating a lack of clear natural boundaries for defining disorder presence or absence (many exist on a continuum), and (2) substantial heterogeneity within each condition. In this regard, heterogeneity includes the issue that different neural network routines may drive diagnosis for different subsets of individuals. In addition, biological measures may differ for one subtype, but not another. As a result, some biological markers may only be found within a specific subset of individuals for a given diagnosis but not in others.

Attention Deficit Hyperactivity Disorder-Inattentive Type as a Distinct Entity

There is a good deal of research to support the position that ADHD-inattentive type represents a pathology largely separate from ADHD "combined type" or "impulsive type." We have reviewed the research that demonstrates that both shared and unique genetic risk factors underlie the two symptom domains of ADHD-inattentive type and impulsive type (Kunsti et al., 2014). We have also shown that the developmental course and relationship of these two disorders differ across their respective symptom domains. This emphasizes the importance of their partially distinct etiologies and the possibility that they represent two different disorders. Research has indicated that two core cognitive impairments phenotypically associated with ADHD symptoms, represented by Reaction Time Variability and Commission Errors on neuropsychological instruments, showed different genetic relationships to the two ADHD symptom domains, suggesting separate neural pathways underlying inattention and hyperactivity–impulsivity symptoms. More recent research demonstrated that the ADHD-impulsive/hyperactive type is distinguishable from the ADHD-inattentive type and ADHD-combined type groups. The impulsive type characterized by fewer externalizing behaviors, poorer sustained attention, and better response inhibition function (Wu et al., 2023). Again, the combining of two pathologies needs to be addressed as it implicates diagnosis, treatment, and research.

In addition to the above, research clearly suggests that there are significant functional connectivity differences between the inattentive and other subtypes of ADHD (Lee et al., 2021). They used diffusion and resting-state functional magnetic resonance imaging data obtained from the Healthy Brain Network database. With this data, they assessed the structural and functional neural network efficiency and structure–functional network (SC–FC) coupling (functional connectivity between network components) at the default mode (DMN), executive control (ECN), and salience (SAN) intrinsic networks of children diagnosed with the inattentive subtype (ADHD-I), the combined subtype (ADHD-C), and typically developing children (TDC). The goal was to compare ADHD symptoms of the existing ADHD subgroups to those of TDC and also assess for any differences between ADHD subtypes. Their finding indicated that, when compared to typically developing children, children with ADHD had lower structural connectivity and network efficiency in the DMN, without significant group differences in functional networks. There were, however, differences. Children with ADHD-C had higher SC–FC coupling, a finding associated with reduced cognitive flexibility, for all subnetworks compared to TDC. The ADHD-C group also demonstrated increased SC–FC coupling in the DMN compared to the ADHD-I group. The correlation between SC–FC coupling and hyperactivity scores was negative in the ADHD-I, but not in the ADHD-C group. The posited ADHD-C and ADHD-I may differ with respect to their underlying neuronal connectivity and that the added dimensionality of hyperactivity may not explain this distinction. Earlier work indicated presaged this finding (Solanto et al., 2009). They found that children with ADHD-combined type and ADHD-inattentive type activated overlapping regions of right inferior frontal gyrus, right superior temporal lobe, and left inferior parietal lobe during inhibitory control. What was different between these groups was the magnitude of the activation in the temporal and parietal regions, as well as in the bilateral middle frontal gyrus. Specifically, the magnitude of activation was greater in children with ADHD-inattentive type than those with ADHD-combined type. In addition, children with ADHD-combined type activated bilateral medial occipital lobe to a greater extent than children with ADHD-inattentive type. The authors concluded that their results provided preliminary evidence that phenotypic differences between the ADHD-C and ADHD-PI subtypes are associated with differential activation of brain regions that have previously been implicated in the pathophysiology of ADHD and are thought to mediate executive and attentional processes.

The idea that ADHD-impulsive type and ADHD-inattentive type are distinct disorders is not new provided a model wherein the core problem in the inattentive type of ADHD is deficits in working memory. She also posited other etiologies such as children with the truly inattentive type of ADHD, rather than being distractible, may instead be easily bored, with their problem being more in motivation (under arousal) than in inhibitory control. We have suggested as much in the chapter on the reward recognition system in this book. As to the underlying network properties of the proposed disorder, Diamond points to the converging evidence of a primary disturbance in the striatum (a frontal–striatal loop) in the combined type of ADHD and proposed that the primary disturbance in truly inattentive-type ADHD (ADD) is in the cortex (a frontal–parietal loop). This would be consistent with the top-down modeling associated with the reward recognition system. Diamond hypothesized that these are not two different types of ADHD, but two different disorders with different cognitive and behavioral profiles, different patterns of comorbidities, different responses to medication, and different underlying neurobiology.

In 2006, Milich, Ballentine, and Lynam reviewed research assessing differences in attention deficit hyperactivity disorder (ADHD) subtypes to assess for the possibility that ADHD-combined type (ADHD/C and ADHD/predominantly inattentive type (ADHD/I)) are distinct and unrelated disorders. They looked for differences between subtypes. Subtypes were assessed along dimensions that included essential and associated features, demographics, measures of cognitive and neuropsychological functioning, family history, treatment response, and prognosis. Important differences among subtypes were indicated in several of these areas, which they concluded as supporting the idea that ADHD/C and ADHD/I may best be characterized as distinct disorders.

While we recognize that arguing for a separate diagnosis represents a continuation of a long-standing, and as yet unresolved, debate, we wish to point out that there are clear benefits to recognizing ADHD-inattentive presentation as a separate disorder. Foremost among these is the idea that lumping all the various types of ADHD presentations together obscures important etiological, in general, and neural network differences in particular. This means that important diagnostic, clinical, and treatment implications may be lost due to being drowned out by the negative features associated with heterogeneous sampling (Bitler et al., 2006). As an example, we can refer back to data suggesting differential impact on overall cognition and language as a result of subtypes. As we have seen, important potential targeted treatment effects are potentially lost. Specifically, recognizing the top-down nature of ADHD-inattentive type implies that issues of salience and reward value will be greatly highlighted with the likelihood that new interventions will be developed (Edel et al., 2013; Pfiffner et al., 2007).

Attentional functioning (and poor functioning) could provide a meaningful, and empirically valid, way to understand the difference between traditional disorders of mental health and pave the way for a new nosology.

We have seen that significant attention issues are associated with different disorders of mental health. We have also seen how these attention differences are unpinned by different patterns of neuroconnectivity associated with the different disorders. It is therefore possible to conceptualize a nosology of disorders that would describe different, and newly defined, disorders of mental health based on the difference in attentional functionality (Sinnott-Armstrong et al., 2024). Their model distinguishes between two types of attention failure: a failure of tendencies or a failure of abilities. A failure of tendencies occurs when an individual repeatedly fails to direct or maintain their attention toward things to which they (should) have the most reason to pay attention, or when they repeatedly fail to shift their attention away from those to which they have reason not to pay attention. Think about the child in math class who is looking at items on the teacher's desk rather than listening to the teacher explain how to solve the problem on the board. If asked, the child would likely state that they recognized that the correct thing to attend to was the lecture, but might add that the item on the teacher's desk was just so very interesting, and for some reason important to them.

A failure of abilities occurs when the individual cannot sustain, direct, or shift their attention despite having valid reasons for doing so. This grouping aligns with the standard dysfunction-based model of the DSM-5 (Diagnostic and statistical manual of mental disorders (5th ed.), 2013).

Obviously, their model of the attention underpinning of most mental disorders is based on top- down regulation. Based upon this distinction, it becomes possible to classify mental health disorders into two classes, the understanding of which may provide useful insight into new and effective treatments. It may also be possible to understand why certain clusters of these disorders demonstrate some degree of symptom overlap. They propose the following subdivisions:

Group 1. Failure to sustain attention to a stimulus in the environment when there is reason to pay attention to it.

Behavioral dysfunction	Emotional dysfunction	Disruption of cognition
ADHD	Depression	Delusions

Group 2. Failure to shift attention when there is reason to do so.

Behavioral dysfunction	Emotional dysfunction	Disruption of cognition
Eating disorders	Phobias	Narcissism

We provided the example of an individual with anorexia nervosa. In an attention-based model, a person with this disorder cannot keep their thoughts from dwelling on body shape or weight and the things that might affect those variables. This is because of the distorted perception of themselves and the world around them. The misperception alone is not sufficient to produce the dysfunction associated with anorexia nervosa. This results from the individual's inability to stop the negative thoughts that cause the disorder. In contrast, depression, which falls within the same cluster, results from the individual's inability to focus on the more adaptive and positive life circumstances that an individual encounters. Other research has made a similar argument, highlighting the similarities of attentional dysfunction inherent among different DSM-5-defined disorders such as schizophrenia and autism (Maiese, 2023).

The assumption of these models is that attention underlies the symptomology of various disorders of mental health and that variability in the underlying neural networks of these disorders are diagnostic in their own right. It is even possible that attention control issues may precede emotional dysregulation for some disorders, and that these attention control factors may in fact be etiologic. For example, it is quite possible that attentional predisposition, inflexibility, and rumination may be causal of a future depressive condition rather than caused by the depression (Sanchez-Lopez et al., 2019).

Research Design Criteria Model

Clearly, what is needed is a system to allow emerging research to incorporate genetic, biological, and social factors that contribute to a system that includes symptoms, while teasing apart these symptoms that currently run across what are perceived as discrete entities. We would highlight that this will extend to include underlying, common neural pathways that subserve these disorders and posit that is possible to generate a newer nosology based on neural network *characteristics* as opposed to labeling distinctive but nonspecific behavioral characteristics. A first step in this direction was made with the introduction of the Research Domain Criteria model (Insel, 2014). The model is not for clinical diagnostic purposes, but rather it is attempting to make its mark by beginning with research—cleaning up the terms and clarifying the features that run across current labels. The model provides views of mental health and psychopathology in the context of major domains (dimensions) of basic human neurobehavioral functioning. Rather than utilizing the traditional practice of symptoms and signs for disease state definitions, and then seeking specific pathologies, RDoC reverses the idea by considering psychopathology in terms of deviations from normal functions that could account for particular symptoms. These symptoms can occur on a transdiagnostic basis.

The RDoC currently consists of five general cognitive and motivational domains. It is organized in a matrix where rows represent specific dimensions of function (constructs) and columns represent areas for study (units of analysis). A construct (e.g., "working memory") represents a particular functional aspect of behavior, and it is tied to knowledge about underlying neurobiological and genetic mechanisms. Constructs are grouped together to form "domains".

Five major functional domains characterize various types of neurobehavioral functioning in the RDoC. Different aspects of each domain are represented by psychological/biological dimensions, or constructs, which are considered along the full range of functioning from normal to abnormal. The model is based on the idea that both behavioral and biological aspects of functioning change and mature across the life span. Equally important is the study of various aspects of the environment, including the physical environment, cultural components, and factors such as social determinants of health. Researchers are encouraged to measure and integrate multiple classes of data (units of analysis, e.g., behavior, genetics, physiology, and self-report data) to seek a comprehensive understanding of the construct(s) under study. Research designs with such an integrative approach are essential in order to obtain a comprehensive understanding both of normal functioning and psychopathology. The Research Domain Criteria classifies mental disorders based on these six domains of human functioning and behavior:

Positive valance system (PVS) and negative valence system (NVS):

PVS is responsible for actions and responses related to motivation, reward, and pleasure. In contrast, NVS drives responses in aversive contexts including loss and threat.

Cognitive systems:

The cognitive systems domain of the RDoC system is a collection of cognitive functions, including attention, declarative memory, language, perception, working memory, and cognitive control.

Systems for social processes:

The social processes domain of the RDoC framework is a set of constructs that represent the systems that underlie interpersonal interactions. Also included are the behavioral and neurobiological processes that promote social relationships and positive social interactions.

Arousal/regulatory systems (attention included):

The arousal and regulatory systems domain of the RDoC framework includes constructs that are responsible for activating and regulating neural systems, energy balance, and sleep.

Sensorimotor systems:

Sensory motor processes are a domain in the RDoC framework of the National Institute of Mental Health (NIMH). The domain includes four constructs:

- Motor action: Includes subconstructs such as action planning and selection, sensorimotor dynamics, initiation, execution, inhibition, and termination
- Agency and ownership
- Habit
- Innate motor patterns

Disorders of mental health as a spectrum:

Another way to consider disorders of mental health and the neural systems that subserve them is to consider the idea of dimensionality. Dimensionality is the idea these disorders occur on a spectrum and represent a continuous range of functioning. To this idea we would add that this range of function reflects the efficiency of the neural networks that subserve the various dimensions. The dimensionality approach reflects a reaction to the ongoing inability of research to find any genetic or neuroscientific evidence to support the breakdown of complex mental disorders into discrete separate categories.

Mental disorder is not an all-or-none phenomenon (Clark et al., 2017). Any mental disorder is continuously scaled in severity, from absence to severe psychopathology. The degree of severity of mental illness is one of its most important aspects, and its dimensional nature is not a problem conceptually. Psychological science uses a number of artificially designated categories as low average, average, and above average that represent artificially but useful divisions along the continuum of intellectual abilities.

The idea of dimensionality of mental disorders is largely at odds with the existing categorical systems such as the DSM-5 or the ICD. While some diagnoses carry a dimensional aspect, such as in distinguishing major depression from that of persistent depressive disorder (dysthymia), in these categorical systems, each entity is distinct and appears only once in a clearly defined place, with emphasis on it being present or not. With a few exceptions, ICD and DSM both classify mental disorders as if they were distinct entities, even while acknowledging that people diagnosed with different disorders often have various characteristics in common. "Moreover, they do so despite the fact that we know that mental disorders are not distinct disorders but complex combinations of psychological problems, which themselves are dimensional. Clearly, reconciling the complex, multidimensional nature of mental illness with the structure of these classification systems is a major challenge" (Clark et al., 2017).

As we have demonstrated, many types of disorders of mental health share significant aspects of foundational processes such as attention. According to this dimensional view, disorders of mental health are the product of shared risk factors that lead to abnormalities in intersecting drives such as motivation and reward anticipation, which can be measured (hence "dimension") and used to place people on one of several spectra (Adam, 2013). Cradock and Owen (2010) proposed such a dimensionality model. While we, in this volume, stress the dimensionality of attention as reflected in many disorders, Cradock and Owen used available data from genetic studies on the overlap of the genetic foundation of schizophrenia and bipolar disorder to underpin the argument for their model. Genome-wide association studies (GWAS) have demonstrated the existence of common DNA variants (single nucleotide polymorphisms) that influence the risk of both schizophrenia and bipolar disorder. They note that there is substantial genetic overlap between schizophrenia and bipolar disorder as evidenced from a large-scale study of bipolar disorder and schizophrenia in which thousands of individuals were studied for hundreds of thousands of common DNA variants spread across the genome. They identify research indicating that structural genomic variants of small or modest size (copy number variants (CNVs) of stretches of DNA of 1000 base pairs or more) are a common cause of genetic variation in humans, and these variants have been reported in neuropsychiatric phenotypes, including autism, intellectual disability, and schizophrenia. Data suggests that the overall "load" of CNVs has been shown to be greater in individuals with schizophrenia compared with controls. This data reflected the shared genetics of the disorders and did not indicate which functional aspects of neurocognitive functioning were impacted with the genetic variants. Such work remains to be done.

There is epidemiological evidence that disorders of mental health share common characteristics and perhaps common primary network underpinning. In their study involving more than five million subjects, they found that comorbidity among various mental disorders was pervasive. Their data indicated that there was an increased risk of developing all other mental disorders after an index mental disorder and that the risk of developing comorbidity was most prominent in the first year after the onset of the initial indexed mental disorder. In addition, the increased risk of developing additional disorders persisted over at least 15 years. For some disorders (e.g., mood disorders), the absolute risks of developing specific later disorders (e.g., anxiety disorders) was substantial (e.g., 30–40% over 5 years). They suggest that "patterns of comorbidity within mental disorders may also reflect a shared risk architecture underlying different types of disorders." By extension, this data points to underlying commonalities between disorders and suggests that there are precursor indices for the development of disorders. There is, in fact, a significant line of research that suggests the existence of a broad, general-psychopathology dimension (Clark et al., 2017) that encompasses a wide range of, perhaps even all, variations of psychopathology, perhaps in a very fundamental way. As we have indicated, disrupted attention in either application or focus is a likely candidate for such a precursor factor. For example, attentional control, which can range from distractibility to hyper attentiveness, can cause problems found in a multiplicity of disorders.

Recently, the HiTOP model was developed to further expand on this idea of dimensionality. They indicate that "Symptom components and maladaptive traits can be thought of as the constituent parts of broader mental health syndromes. For example, social anxiety is a syndrome that brings together various expressions of performance and social interaction anxiety. Syndromes are conceptualized as dimensional tendencies to experience related symptoms and maladaptive traits. In other words, they are continuously distributed concepts, not categories, that span a range of severity" (Fig. 12.1).

While this approach emphasizes functional emotional states or specific behaviors, we believe the same can be done for foundational neural network functions such as attention and provided the Sinnott-Armstrong, Soter, and Summers (2024) model as an example.

Neurobehavioral Functioning

Other diagnostic markers based on neural connectivity and regional specialization of function have been put forward. Described an intermediate scale analysis that characterized temporal variability of the functional architecture associated with a particular brain region. Temporal variability (TV) refers to measures of how the functional connectivity of a brain region changes over time. It can reflect how brain regions reconfigure into different functional modules at different times. TV analysis can capture the full extent of resting-state activities, which can reflect cognitive functions and rehabilitation potential. Zang et al. developed a topography of variability that reflected the whole-brain and created an analytical framework to establish the fundamental relationship between variability of regional functional architecture and its neural activity or structural connectivity. Their results indicated that a region's temporal variability reflected the dynamic reconfiguration of a brain

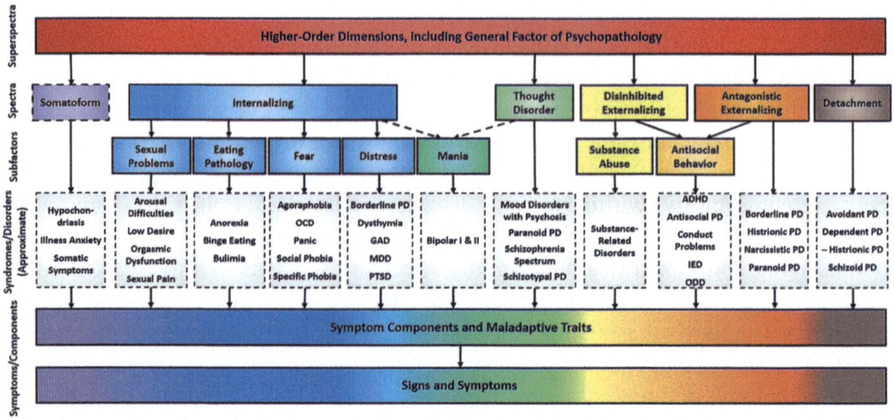

Fig. 12.1 HiTOP model

region into distinct functional modules at different times and suggest this may be indicative of brain flexibility and adaptability. In their study, combinations of adaptability and flexibility characterized different disorders of mental health. Specifically, primary and unimodal sensory-motor cortices demonstrated low temporal variability, while transmodal areas, including heteromodal association areas and limbic system areas, demonstrate the high variability. Transmodal regions integrate information from multiple functional networks and play an important role in cognition. They are also known as nonsensory-specific regions and are located between primary brain areas, such as the somatosensory, motor, auditory, and visual areas, and other transmodal areas, such as the prefrontal, temporal, and inferior parietal cortex. Transmodal areas include the heteromodal cortex, paralimbic cortex, and limbic cortex. Heteromodal areas receive input from all sensory modalities or transmit data to multiple secondary motor cortices. They are also known as the heteromodal association cortex (HASC) or tertiary cortex. Given this understanding, the findings of combinations of areas and temporal variability resulting in different disorders makes sense.

In particular, regions with highest variability such as hippocampus/parahippocampus, inferior and middle temporal gyrus, olfactory gyrus, and caudate are all related to learning, suggesting that the temporal variability may indicate the level of brain adaptability. For example, their research identified completely opposing patterns of variability changes between schizophrenia and attention deficit hyperactivity disorder/autism. Specifically, regions of the default-mode network demonstrated lower variability in individuals with schizophrenia, but high variability in patients with autism/attention deficit hyperactivity disorder, compared with respective controls. In contrast, subcortical regions, particularly the thalamus, displayed higher variability in schizophrenia patients, but lower variability in patients with attention deficit hyperactivity disorder. The changes in variability of these regions were highly associated to symptom scores.

The point here is that it is possible to create new nosologies that will drive research with the potential to yield better understanding of these disorders and the potential to uncover new and more effective treatments. It is highly likely that attention variability will be a key component in these developing systems.

Finally, there is the idea that in many instances, attention is not disordered at all, it is merely misapplied or less than maximally efficiently utilized.

There are disorders where attention seems disrupted because the impacted individual experiences difficulty attending to that which they are supposed to be attending. We have seen that in some instances this apparent attention problem does not represent a dysfunction in the attention system but rather a misapplication of it. These are apparent attention problems on a behavioral level, but may require a different treatment than that offered for attention deficit hyperactivity disorder. We have looked at examples wherein the "umbrella" diagnosis is ADHD and we would expect a disorder in attention. Does the premise hold for other diagnostic labels?

For example, let us look at what is known as the discrete diagnostic label of obsessive-compulsive disorder (OCD). Levy (2018) points out that obsessive–compulsive disorder is traditionally thought of as caused (in large part) by distinctive personality traits and belief biases. The research, however, indicates that a substantial number of impacted individuals (bearing this diagnosis) do not manifest these traits. To replace the traditional ideas of causality, Levy proposed a predictive coding explanation of the disorder, which explained both the symptoms and the cognitive traits. Predictive coding is a theory in neuroscience that proposes the brain constantly generates and updates a mental model of its environment, which it uses to predict sensory input signals. The brain then compares these predictions to the actual sensory input signals. This process is also known as predictive processing, prediction error minimization, or free energy minimization. According to this model, attention becomes a central player. Specifically, OCD centrally involves heightened and dysfunctionally misapplied attention to normally unattended sensory and motor representations. As these representations have contents that predict catastrophic outcomes, impacted individuals are disposed to engage in behaviors and mental rituals designed to block these outcomes or to produce more precise information. The same representations also cause the cognitive traits characteristic of impacted individuals.

We have also seen how attention is misapplied rather than disordered in bipolar disorder. In one of the earliest studies on this aspect of attention, Jongen, Smulders, Ranson, Arts, and Krabbendam (2007) found that mildly depressed clients focused their attention away from depression-related and positive words and reported that this finding was mood state-dependent. They also found that the bias away from positive words could be seen in euthymia, which may be part of a trait effect. Interestingly, individuals presenting with mania were found to show no attentional vigilance to happy or sad expressions, but they avoided sad expressions and focused more on happy expressions in later emotion processing. This finding suggests distinct attentional biases may be present in both depressive and manic episodes. Similar misapplication issues are also likely associated with other disorders, such as addictive gambling for example (Duller et al., 2021).

The extant research suggests that ADHD is extensively comorbid with many mood disorders (Mao & Findling, 2014; Sandstrom et al., 2021). Their results indicated that there is a comorbidity rate of up to 75% between ADHD and bipolar disorder and a comorbidity of up to 25% in major depressive disorder. While we do not argue that the identified individuals met the behaviorally defined criteria of a diagnosis of attention disorder, we would posit that at least some of the individuals so diagnosed were misapplying their attention as opposed to having it disrupted. The fact is we cannot know because the lumping together of all sorts of attentional dysfunction under one broad heterogeneous category obscures us from knowing. This is the very problem that a more thorough understanding of all forms of attentional deficiency would prevent. Understanding these dysfunctional aspects of attention would inevitably lead to better research and better treatment.

The multiplicity of symptom patterns and potential causes for a particular disorder is referred to as etiologic heterogeneity. Etiologic heterogeneity describes a situation where different subtypes of a disorder are shaped by different risk factors (Girault et al., 2024). We have described just this type of situation when it comes to the multiplicity of disorder or misapplied attention we have described in this book. The various subtypes can be differentiated, and classified, by the level of severity, genetic markers, comorbidities, or many other dimensions. They are potentially established as being causally distinct when differentially influenced by one or a cluster of etiologic factors.

The are a number of empirical reasons why we should consider etiologic heterogeneity and several valid methodologies to do so (Kalkbrenner et al., 2024). Incorporating etiologic heterogeneity optimizes the goals of improving health in a number of ways. Observing that one subgroup is more (or less) impacted by a hypothesized causal factor, whether an environmental agent or a genetic variant, provides knowledge about shared versus distinct causal pathways. These factors can be combined with other factors, such as clinical and phenotypic dimensions of the subtype, and pathophysiologic and toxicologic information about causal risk factors that would allow for the construction of more complete pictures of disease etiology. As we have seen, another advantage is enhanced statistical power, emanating from isolating a more etiologically homogeneous subgroup that potentially is more responsive to a given risk factor leading to a stronger measure of association, which would have been diluted had that subgroup been combined with nonresponsive subgroups (Begg & Zabor, 2012). Given that various subtypes may contain distinctions in etiology on the basis of known risk factors, a legitimate question arises as to whether undiscovered risk factors may be more or less likely to have distinct risks for the subtypes. They note that "as the credibility of the hypothesis that undiscovered risk factors may influence only one subtype or may have a differential effect on different subtypes becomes more plausible, the strategy of searching for new risk factors separately within subtypes becomes a more attractive one." Lastly, better-targeted interventions may follow from identifying distinct subtypes.

Mental health, psychiatric, and neurodevelopmental disorders especially benefit from studies of etiologic heterogeneity. As we have noted, most of these conditions are highly heterogeneous with behavioral symptom patterns that overlap across multiple behaviorally defined disorders. As we have also noted, the search for more homogenous subtypes, coupled with new classifications to advance the understanding and treatment of mental health disorders, is increasingly recognized as important. It is evidenced by searches for endophenotypes, efforts to subclassify based on genetic markers, and the RDoC initiative of the National Institute of Mental Health. As attention represents a foundational, inherent skill set present from birth and involved in the complexities of neuromaturation and neurodevelopment, the effort should include a focus to include attention disorders, as well as the multiplicity of subtypes associated with it.

It is a construct with a broad scope of implications. It is a construct that must be appreciated in its fullness in order to understand the implications for clinical practice and research.

References

Abramowitz, J. S., Taylor, S., & McKay, D. (2009). Obsessive-compulsive disorder. *The Lancet, 374*(9688), 491–499.

Achenbach, T. (1991). *Manual for the child behavior checklist/4–18 and 1991 profile*. Department of Psychiatry, University of Vermont.

Adam, D. (2013). Mental health: On the spectrum. *Nature, 496*, 416–418. https://doi.org/10.1038/496416a

American Psychiatric Association. (2013). *The diagnostic and statistical manual of mental disorders* (5th ed.). DSM–5, American Psychiatric Association.

Anguera, J. A., Gunning, F. M., & Areán, P. A. (2017). Improving late life depression and cognitive control through the use of therapeutic video game technology: A proof-of-concept randomized trial. *Depression and Anxiety, 34*(6), 508–517.

Barkley, R., & Fischer, M. (2010). The unique contribution of emotional impulsiveness to impairment in major life activities in hyperactive children as adults. *Journal of the American Academy of Child and Adolescent Psychiatry, 49*(5), 503–513.

Begg, C., & Zabor, E. (2012). Detecting and exploiting etiologic heterogeneity in epidemiologic studies. *American Journal Epidemiology, 176*, 512–518. https://doi.org/10.1093/aje/kws128

Bitler, M., Gelbach, J., & Hoynes, H. (2006). What mean impacts miss: Distributional effects of welfare reform experiments. *American Economic Review, 96*(4), 998–1012. https://doi.org/10.1257/aer.96.4.988

Chen, Z., He, C., Zhang, P., Cai, X., Li, X., Huang, W., ... & Zhang, Y. (2024). Brain network centrality and connectivity are associated with clinical subtypes and disease progression in Parkinson's disease. *Brain Imaging and Behavior*, 1–16.

Clark, L., & Goodwin, G. M. (2004). State-and trait-related deficits in sustained attention in bipolar disorder. *European Archives of Psychiatry and Clinical Neuroscience, 254*, 61–68.

Clark, L., Cuthbert, B., Lewis-Fernandez, R., Narrow, W., & Reed, G. (2017). Three approaches to understanding and classifying mental disorder: ICD-11, DSM-5, and the national institute of mental health's research domain criteria (RDoC). *The Public Interest, 18*(2), 72. https://doi.org/10.1177/15291006177272

Coghill, D., & Banaschewski, T. (2009). The genetics of attention-deficit/hyperactivity disorder. *Expert Review of Neurotherapeutics, 9*(10), 1547–1565. https://doi.org/10.1586/ern.09.78

Connors, K. (2022). Parent rating scale. In *(Conners 4™) Conners* (4th ed.). WPS Publishing.

Costa Dias, T., Wilson, V., Bathula, D., Iyer, S., Mills, K., Thurlow, B., et al. (2013). Reward circuit connectivity relates to delay discounting in children with attention-deficit/hyperactivity disorder. *European Neuropsychopharmacology, 23*, 33. https://doi.org/10.1016/j.euroneuro.2012.10.015

Dias, T. G. C., Iyer, S. P., Carpenter, S. D., Cary, R. P., Wilson, V. B., Mitchell, S. H., ... & Fair, D. A. (2015). Characterizing heterogeneity in children with and without ADHD based on reward system connectivity. *Developmental Cognitive Neuroscience, 11*, 155–174.

Duller, P., Krishnan, V., & Diaz, A. (2021). A systematic review on the intersection of attention-deficit hyperactivity disorder and gaming disorder. *Journal of Psychiatric Research, 133*, 212–222. https://doi.org/10.1016/j.jpsychires.2020.12.026

Edel, M.-A., Enzi, B., Whitthaus, H., Tegenthoff, M., Peters, S., Juckel, G., & Lissek, S. (2013). Differential reward processing in subtypes of adult attention deficit hyperactivity disorder. *Journal of Psychiatric Research, 47*(3), 350–356. https://doi.org/10.1016/j.jpsychires.2012.09.026

Feczko, E., Miranda-Dominguez, O., Marr, M., Graham, A., Nigg, A., & Faira, D. (2019). The heterogeneity problem: Approaches to identify psychiatric subtypes. *Trends in Cognitive Science Jul, 23*(7), 584–601. https://doi.org/10.1016/j.tics.2019.03.009

Gioia, G., Isquith, P., Retzlaff, P., & Espy, K. (2002). Confirmatory factor analysis of the behavior rating inventory of executive function (BRIEF) in a clinical sample. *Child Neuropsychology, 8*(4), 249–257. https://doi.org/10.1076/chin.8.4.2

Girault, J., Veatch, O., & Won, H. (2024). Etiologic heterogeneity, pleiotropy, and polygenicity in behaviorally defined intellectual and developmental disabilities. *Journal of Neurodevelopmental Disorders, 16*, 8. https://doi.org/10.1186/s11689-024-09526-z

Graziano, P., & Garcia, A. (2016). Attention-deficit hyperactivity disorder and children's emotion dysregulation: A meta-analysis. *Clinical Psychology Review, 46*, 106–123. https://doi.org/10.1016/j.cpr.2016.04.011

Insel, T. (2014). The NIMH research domain criteria (RDoC) project: Precision medicine for psychiatry. *The American Journal of Psychiatry, 171*(4), 395–397. https://doi.org/10.1176/appi.ajp.2014.14020138

Kalkbrenner, A., Zheng, C., Yu, J., Jenson, T., Kuhlwein, T., Ladd-Acosta, C., & Schendel, D. (2024). Method for testing etiologic heterogeneity among noncompeting diagnoses, applied to impact of perinatal exposures on autism and attention deficit hyperactivity disorder. *Epidemiology, 35*(5), 689–700. https://doi.org/10.1097/EDE.0000000000001760

Keller, A., et al. (2019). Paying attention to attention in depression. *Translational Psychiatry, 9*, 279. https://doi.org/10.1038/s41398-019-0616-1

Kunsti, J., et al. (2014). The separation of ADHD inattention and hyperactivity-impulsivity symptoms: pathways from genetic effects to cognitive impairments and symptoms. *Journal of Abnormal Child Psychology, 12*, 127–136. https://doi.org/10.1007/s10802-013-9771-7

LaBianca, S., Brikell, I., Helenius, D., Loughnan, R., Mefford, J., Palmer, C. E., ... & Schork, A. J. (2024). Polygenic profiles define aspects of clinical heterogeneity in attention deficit hyperactivity disorder. *Nature Genetics, 56*(2), 234–244.

Lee, D., Quattrocki Knight, E., Song, H., Lee, S., Pae, C., Yoo, S., & Park, H. (2021). Differential structure-function network coupling in the inattentive and combined types of attention deficit hyperactivity disorder. *PLoS One, 16*(12), e0260295. https://doi.org/10.1371/journal.pone.0260295

Levy, N. (2018). Obsessive–compulsive disorder as a disorder of attention. *Mind and Language, 33*(1), 3–16. https://doi.org/10.1111/mila.12172

Luo, Y., Weibman, D., Halperin, J., & Li, X. (2019). A review of heterogeneity in attention deficit/hyperactivity disorder (ADHD). *Frontiers in Human Neuroscience, 13*, 42. https://doi.org/10.3389/fnhum.2019.00042

Maiese, M. (2023). Are all mental disorders affective disorders? *Passion: Journal of the European Philosophical Society for the Study of Emotions, 1*(1), 31–49. https://doi.org/10.59123/passion.v1i1.12728

Mao, A., & Findling, R. (2014). Comorbidities in adult attention-deficit/hyperactivity disorder: A practical guide to diagnosis in primary care. *Postgraduate Medicine, 126*(5), 42–51. https://doi.org/10.3810/pgm.2014.09.2799

Nigg, J., & Casey, B. (2005). An integrative theory of attention-deficit/hyperactivity disorder based on the cognitive and affective neurosciences. *Development and Psychopathology, 17*, 785–806. https://doi.org/10.1017/s0954579405050376

Pfiffner, L., Mikami, A., Huang-Pollack, C., Easterlin, B., Zalecki, C., & McBurnett, K. (2007). A randomized, controlled trial of integrated home-school behavioral treatment for ADHD, predominantly inattentive type. *Journal of the American Academy of Child & Adolescent Psychiatry, 46*(8), 1041–1050. https://doi.org/10.1097/chi.0b013e318064675f

Reale, L., Bartoli, B., Cartabia, M., Zanetti, M., Costantino, M., C. M., et al. (2017). Comorbidity prevalence and treatment outcome in children and adolescents with ADHD. *European Child and Adolescent Psychiatry, 26*, 1443–1457. https://doi.org/10.1007/s00787-017

Roberts, C. S. (2023). *Impact and centrality of attention dysregulation on cognition, anxiety and low mood in adolescents* (Vol. 13). Scientific Reports.

Salum, G., Gadelha, A., Polanczyk, G. M., & Rohde, L. (2018). Diagnostic operationalization and phenomenological heterogeneity in psychiatry: the case of attention deficit hyperactivity disorder. *Salud Ment, 41 no.6*, 249. https://doi.org/10.17711/sm.0185-3325.2018.037

Sanchez-Lopez, A., Koster, E., & Van Put, J. (2019). Attentional disengagement from emotional information predicts future depression via changes in ruminative brooding: A five-month longitudinal eye-tracking study. *Behavior Research and Therapy, 118*, 30–42. https://doi.org/10.1016/j.brat.2019.03.013. Epub 2019 Mar 27.

Sandstrom, A., Perroud, N., Alda, M., Uher, R., & Pavlova, B. (2021). Prevalence of attention-deficit/hyperactivity disorder in people with mood disorders: A systematic review and meta-analysis. *Acta Neuropsychiatrica Scandinavica, 143*(5), 380–391. https://doi.org/10.1111/acps.13283

Sinnott-Armstrong, W., Soter, L., & Summers, J. (2024). Mental disorders as failsures of attention. *Crítica. Revista Hispanoamericana De Filosofía., 17-44*, 17. https://doi.org/10.22201/iifs.18704905e.2024.1528

Solanto, M., Shuktz, K., Fan, J., Tang, C., & Newcorn, J. (2009). Event-related fmri of inhibitory control in the predominantly inattentive and combined subtypes of ADHD. *Journal of Neuroimaging, 19*, 205. https://doi.org/10.1111/j.1552-6569.2008.00289.x

Thapar, A., Langley, K., O'Donovan, M., & Owen, M. (2006). Refining the attention deficit hyperactivity disorder phenotype for molecular genetic studies. *Molecular Psychiatry, 11*(8), 714–720.

Wang, X., Zhou, W., & Zhu, X. (2020). Attention deficits in adults with Major depressive disorder: A systematic review and meta-analysis. *Asian Journal of Psychiatry, 53*(October), 102359. https://doi.org/10.1016/j.ajp.2020.102359

Weiss, M. (2000). *Weiss functional impairment rating scale (WFIRS) self-report*. University of British Columbia.

Wu, Z., Wang, P., Liu, J., Liu, L., Cao, X., Sun, L., & Yang, B. (2023). The clinical, neuropsychological, and brain functional characteristics of the ADHD restrictive inattentive presentation. *Frontiers in Psychiatry, 14*, 1099882. https://doi.org/10.3389/fpsyt.2023.1099882

Appendix: Types of Attentional Dysfunction

For the purpose of diagnoses, the current clinical practice has the clinician looking at problematic behavior and the attempting to match the observed clinical features to preestablished behaviorally defined clinical categories of dysfunction. This process is circuitous and reinforces the current nosology. It does not encourage the development of newer models and specifically does not reflect the functioning of the neural network organization of the human brain. As our understanding of the operation of the neural network increased, so does the need to reimagine the ways in which that network process may become dysfunctional.

Throughout this book we have, based on neural network modeling, alluded to the many ways that attention can be disrupted. It is a far-ranging list that transcends the typical attentional deficiencies associated with attention deficit hyperactivity disorder. We believe that many of these represent deficiencies in their own right and deserve both clinical and research attention. They are listed here in no particular order of importance or preference.

1. Expectancy Theory posits that low-level stimulus control of attention is distinguishable from high-level cognitive control of attention. Stimulus-driven attention refers to a transient bottom-up involuntary process. Attentional capture exemplifies this sort of low-level attentional control because it typically involves an immediate reaction to a salient or sudden and unexpected stimulus change (novel) that captures attention. Research has indicated that people differ in whether arousal is usually pleasant or unpleasant (Kuppens, 2006). These differences were found to be related to dispositional reward responsiveness. People with low arousal would orient less to novel stimuli and therefore fail to attend to some stimuli that other people may attend to.
2. High-level cognitive control of attention has been associated with voluntary top-down executive processes that guide attending toward a goal over the long term. Expectancy Theory also focuses on the controlling aspects of top-down

T. Wasserman, L. D. Wasserman, *Disorders of Attention*, Neural Network Model: Applications and Implications, https://doi.org/10.1007/978-3-031-78732-4

systems on the regulation of attention. Stimuli are not, for the most part, universally regulated. They differ from individual to individual based on the history of reward associated with them. In some situations, an individual may find a certain stimulus more rewarding than other stimuli and direct their attention to their preferred stimuli as opposed to the stimuli preferred by another. Simply put, a child may find the butterfly outside the window far more interesting than the math problem on the board that the teacher wants attended to. This would appear, on a behavioral level, to represent an attentional problem based on an inability to focus.

3. There is the type of attention that orients a person toward a new stimulus in the environmental field. This is largely related to bottom-up environmental changes based on novelty and uniqueness. Under certain conditions, bottom-up environmental information can overwhelm the central executive and capture attention. The result is inattention to the desired target stimuli caused by elevated distractibility (Oh look, there is a butterfly outside my classroom window). This, of course, can happen in two ways. The first is that there is a plethora of environmental information that exceeds the capacity of the central executive to manage the information flow, and the second, more likely explanation, is that the central executive in that person is inefficient. There is research support for both possibilities.
4. Attentional problems may be related to deficits in executive functioning (Barkley, 2022). In fact, we believe that we have demonstrated this idea very clearly. Specifically, the behavioral dysregulation found in ADHD may be a consequence of deficits in self-monitoring or adaptive control, both of which are required for adaptive behavior. Processing of contextual demands, ongoing monitoring of one's behavior to evaluate whether it is appropriate for a particular situation, and adjusting behavior when it is suboptimal are components of self-regulation (Shiels & Hawk Jr., 2010).
5. There are attentional processes devoted to focusing and sustaining. Problems can arise in both of these areas. For the attentional processes associated with sustaining and focusing, issues related to reward discounting and probabilistic reward determination will be related to types of dysfunctions in the operation of the attentional system. For example, alterations in reward circuits are hypothesized as one core dysfunction in ADHD, leading to altered processing of anticipated rewards. These deficits are associated with nucleus accumbens (NAcc), which is particularly important for reward processes; task-based fMRI studies have found atypical activation of this region while the participants performed a reward task (Costa-Dias et al., 2013).
6. Genetic studies in general and molecular studies in specific have identified numerous candidate genes that contribute to a common outcome disorder we label ADHD. These same studies have suggested dominant and additive genetic etiology. This suggests a cluster of disorders with a similar behavioral outcome. Why are these differing disorders studied together when studying them individually might produce new and more specific treatment interventions?
7. Deficits in attention have been identified in obsessive-compulsive disorder (OCD) (Posner et al., 2017). In OCD, functional connectivity between two

neural networks associated with attentional processes: the default mode network (DMN) and the salience network (SN) are altered. In OCD, there is significantly reduced inverse connectivity between the anterior medial prefrontal cortex (amPFC) and the anterior insular cortex, regions within the DMN and SN, respectively. In OCD, reduced inverse DMN–SN connectivity has been associated with both increased OCD symptom severity and decreased sustained attention. OCD symptoms have been identified with an imbalance in DMN–SN networks that subserve attentional processes, and this effect of OCD on DMN–SN connectivity is associated with decreased sustained attention. This research adds to a growing literature (Fitzgerald et al., 2010; Beucke et al., 2014), suggesting that reduced inverse DMN–SN connectivity may represent a transdiagnostic marker of attentional processes and suggests a potential mechanistic account of the relationship between OCD and attention. SMN and SN connectivity issues have been identified in research on the impact of salience deficits in depressive, and several other, disorders. Collectively this research points to a class of attentional issues unrelated to ADHD.

8. Anxiety significantly impacts the allocation of attention. One way to understand this phenomenon is by utilizing signal detection theory (McNicol, 2004). A signal detection model would posit that individuals learn over time how to calibrate their anxiety responses by learning which cues indicate danger in the environment. The goal is to develop an optimal strategy for doing so. As they develop, individuals are engaged in an ongoing process of exploration, assessment, and exploitation designed to assess the tradeoff between obtaining a better estimate of the level of risk on one hand and maximizing current payoffs on the other (Meacham & Bergstrom, 2016). This assessment and the resulting value of the tradeoff solution is idiosyncratic to the individual and reflects the exigencies of their own life experiences. Meacham and Bergstrom suggest that as a result of this tradeoff, certain people can become trapped in a state of self-perpetuating oversensitivity to threatening stimuli, even when individuals learn optimally. This phenomenon arises because when individuals become too cautious, they stop sampling the environment and fail to correct their misperceptions, whereas when individuals become too careless, they continue to sample the environment and soon discover their mistakes. Thus, oversensitivity to threats becomes common whereas undersensitivity becomes rare. The more oversensitive you are to the threats, the greater your anxiety becomes. Again, remember that the perceived threats can be highly idiosyncratic, and a person may find something in the environment to be highly threatening when the vast majority of their peers would not reach the same conclusion. Thus, the attention system may be working as it should, but the resulting attention is miss/poorly allocated as a result of experience, reinforcement, or value calculations. Attention is disrupted, but in a very different fashion from the disrupted attention caused by ADHD.
9. Other misapplication of attentional resources occurs in depression. Rumination (not shifting attention from a negative stimulus) and a reduced capacity to disengage attention as appropriate to context (attention control deficits) have each

been implicated in the etiology and maintenance of depressive disorders. Results also suggested that the relationship between attention inflexibility and rumination remains independent of depression severity and that these characteristics persisted after the depressive symptoms dissipated. In other words, it is quite possible that attentional predisposition, inflexibility, and rumination may be causal of a future depressive condition and not caused by the depression. The data also suggested that these attentional characteristics remained after treatment and therefore should be targeted in their own right as part of a comprehensive treatment program for depression.

10. We have indicated that a subset of attention deficit hyperactivity disorder-inattentive type represents an inefficiency of an integrated network system designed to allocate working memory to designated tasks rather than the absence or dysfunction of a particular form of attention. This inefficiency in the allocation of working memory represents, in part, poor engagement of the reward recognition network with neural networks subserving learning and performance that control instrumental conditioning (learning). For this subset of individuals who present on a behavioral level with attention deficit hyperactivity disorder-inattentive type, their problems represent the engagement, or lack thereof, of the motivational and reward recognition network as opposed to problems, or disorders of attention traditionally defined as problems with orienting, focusing, and sustaining. We demonstrated that there is an integrated system of working-memory allocation that responds by recruiting relevant aspects of both cortex and subcortex to the demands of the task being encountered. In this model, attention is viewed as a gating function determined by novelty, flight-or-fight response, and reward history/valence affecting motivation (Wasserman & Wasserman, 2015). Subsequent research has substantiated the role of the networks supporting reward recognition for understanding disrupted attention (Anderson, 2017). These networks support the idea that reward history reciprocally guides attention to reward stimuli. One of these networks is the complex and differentiated locus coeruleus norepinephrine (LC–NE) system that is involved with different behavioral and cognitive substrates of reward and attention. Specifically, the LC receives reward-related sensory, perceptual, and visceral inputs, releases NE, glutamate, dopamine, and various neuropeptides, forms reward memories, drives attentional bias, and selects behavioral strategies for reward. This research points to preclinical and clinical studies that have found that abnormalities in the LC–NE system are involved in a variety of psychiatric conditions marked by disturbed functions in reward and attention (Zhang et al., 2023). This line of research understands that conventional models of attention posit a theoretical dichotomy between goal-directed top-down attentional control and stimulus-driven bottom-up attentional control. However, this conventional theory leaves a gap in the understanding of attentional selection and allocation. There is, in fact, a third category, termed "selection history," which indicates that the history of attentional deployments can bias attentional selections (Awh et al., 2012). Specifically, past rewarding experiences detached from current goals and physical salience can posit a lasting effect on attentional control (Anderson, 2019).

References

Abramovitch, A., Dar, R., Hermesh, H., & Schweiger, A. (2011). Comparative neuropsychology of adult obsessive-compulsive disorder and attention deficit/hyperactivity disorder: Implications for a novel executive overload model of OCD. *Journal of Neuropsychology*. https://doi.org/10.1111/j.1748-6653.2011.02021

Achenbach, T. (1991). *Manual for the child behavior checklist/4–18 and 1991 profile*. Department of Psychiatry, University of Vermont.

Adam, D. (2013). Mental health: On the spectrum. *Nature, 496*, 416–418. https://doi.org/10.1038/496416a

Adam, R., Stephan, K., Brown, H., Frith, C., & Friston, K. (2013). The computational anatomy of psychosis. *Frontiers in Psychiatry, 4*, 47. https://doi.org/10.3389/fpsyt.2013.00047

Alderson, M., Kasper, L., Hudec, K., & Patros, C. (2013). Attention-deficit/hyperactivity disorder (ADHD) and working memory in adults: A meta-analytic review. *Neuropsychology, 27*(3), 287–302. https://doi.org/10.1037/a0032371

Alderson, T., Bokde, A., Kelso, J., Mcquire, L., & Coyle, D. (2020). Metastable neural dynamics underlies cognitive performance across multiple behavioural paradigms. *Human Brain Mapping, 41*(12), 3212–3234. https://doi.org/10.1002/hbm.25009

Ambrosini, P., Bennett, D., & Elia, J. (2013). Attention deficit hyperactivity disorder characteristics, II: Clinical correlates of irritable mood. *Journal of Affective Disorders, 145*, 70–76. https://doi.org/10.1016/j.jad.2012.07.014

American Psychiatric Association. (1968). *Diagnostic and statistical manual of mental disorders* (2nd ed.; DSM-II).

American Psychiatric Association. (2013). *The diagnostic and statistical manual of mental disorders* (5th ed.; DSM–5). American Psychiatric Association.

Andersen, B., Lauremt, P., & Yantis, S. (2011). Value-driven attentional capture. *PNAS Biological Science, 108*(25), 10367–10371. https://doi.org/10.1073/pnas.1104047108

Anderson, B. (2011). There is no such thing as attention. *Frontiers in Psychology, 2*, 246. https://doi.org/10.3389/fpsyg.2011.00246

Anderson, B. (2017). Reward processing in the value-driven attention network: Reward signals tracking cue identity and location. *Social Cognitive and Affective Neuroscience, 12*(3), 461–467. https://doi.org/10.1093/scan/nsw141

Anderson, B. A. (2019). Neurobiology of value-driven attention. *Current Opinion in Psychology, 29*, 27–33.

T. Wasserman, L. D. Wasserman, *Disorders of Attention*, Neural Network Model: Applications and Implications, https://doi.org/10.1007/978-3-031-78732-4

Anderson, P., He, X., Buehler, C., Teney, D., Johnson, M., & Gould, S. (2018). Bottom-up and top-down attention for image captioning and visual question answering. In *CVPR proceedings of the IEEE conference on computer vision and pattern recognition* (pp. 6077–6086).

Andoh, J., Matsushita, R., & Zatorre, R. J. (2015). Asymmetric interhemispheric transfer in the auditory network: Evidence from TMS, resting-state fMRI, and diffusion imaging. *Journal of Neuroscience, 35*(43), 14602–14611.

Andrade, M. N. (2023, December 14). *History_of_Attention*. Retrieved from LibreTexts Social Sciences: https://socialsci.libretexts.org/Bookshelves/Psychology/Cognitive_Psychology/Cognitive_Psychology_

Angelopoulou, E., & Drigas, A. (2021). Working memory, attention and their relationship: A theoretical overview. *Research, Society and Development, 10*(5), e46410515288.

Armel, K., Beaumel, A., & Rangel, A. (2006). Biasing simple choices by manipulating relative visual attention. *Judgment and Decision Making, 3*(5), 396–403.

Arnett, A., Rhodes, C., & Rutter, T. (2021). Reduced error recognition explains post-error slowing differences among children with attention deficit hyperactivity disorder. *Journal of the International Neuropsychological Society, 28*(8). https://doi.org/10.1017/S1355617721001065

Arsten, A., & Rubia, K. (2012). Neurobiological circuits regulating attention, cognitive control, motivation, and emotion: Disruptions in neurodevelopmental psychiatric disorders. *Journal of the American Academy of Child and Adolescent Psychiatry, 51*(4), 356–367. https://doi.org/10.1016/j.jaac.2012.01.008

Awh, E., Belopolsky, A., & Theeuwes, J. (2012). Top-down versus bottom-up attentional control: A failed theoretical dichotomy. *Trends in Cognitive Sciences, 16*(8), 437–443. https://doi.org/10.1016/j.tics.2012.06.010

Baddeley, A., & Hitch, G. (1974). Working memory. In G. Bower (Ed.), *Recent advances in learning and motivation* (Vol. 8, pp. 47–49). Academic Press. https://doi.org/10.1016/S0079-7421(08)60452-1

Balkenius, C., Förster, A., Johansson, B., & Thorsteinsdottir, V. (2008). Anticipation in attention. In I. G. Pezzulo, M. Butz, C. Castelfranchi, & R. Falcone (Eds.), *The challenge of anticipation* (Lecture notes in computer science) (Vol. 5225). Springer. https://doi.org/10.1007/978-3-540-87702-8_4

Ball, G., & al, e. (2014). Rich-club organization of the newborn human brain. *Proceedings of the National Academy of Sciences USA, 111*, 7456–7461. https://doi.org/10.1073/pnas.1324118111

Barker, H., Munro, J., Orlov, N., Morgenroth, E., Moser, J., Eysenck, M., & Allen, P. (2018). Worry is associated with inefficient functional activity and connectivity in prefrontal and cingulate cortices during emotional interference. *Brain and Behavior, 8*(12), e01137. https://doi.org/10.1002/brb3.1137

Barkley, R. (2022). Improving clinical diagnosis using the executive functioning—Self-regulation theory of ADHD. *The ADHD Report*. https://doi.org/10.1521/adhd.2022.30.1.1

Barkley, R., & Fischer, M. (2010). The unique contribution of emotional impulsiveness to impairment in major life activities in hyperactive children as adults. *Journal of the American Academy of Child and Adolescent Psychiatry., 49*(5), 503–513.

Barnes, R., & Jones, M. (2000). Expectancy, Attention, and Time. *Cognitive Psychology, 41*(3), 254–311. https://doi.org/10.1006/cogp.2000.0738

Batmaz, S., Altinoz, S., & Sonkurt, H. (2021). Cognitive attentional syndrome and metacognitive beliefs as potential treatment targets for metacognitive therapy in bipolar disorder. *World Journal of Psychiatry, 11*(9), 589–604. https://doi.org/10.5498/wjp.v11.i9.589

Beard, C., Sawyer, A., & Hofmann, S. (2012). Efficacy of attention bias modification using threat and appetitive stimuli: A meta-analytic review. *Behaviour Therapy and Research, 43*(4), 724–740. https://doi.org/10.1016/j.beth.2012.01.002. Epub 2012 Jan 18.

Beck, D., & Kastner, S. (2005). Stimulus context modulates competition in human extrastriate cortex. *Nature Neuroscience, 8*(8), 1110–1116. https://doi.org/10.1038/nn1501

Beck, D., & Krastner, S. (2009). Top-down and bottom-up mechanisms in biasing competition in the human brain. *Vision Research, 49*(10), 1154–1165. https://doi.org/10.1016/j.visres.2008.07.012

Beckmann, C. F. (2005). Investigations into resting-state connectivity using independent component analysis. *Philosophical Transactions of the Royal Society B: Biological Sciences, 360*(1457), 1001–1013.

Begg, C., & Zabor, E. (2012). Detecting and exploiting etiologic heterogeneity in epidemiologic studies. *American Journal Epidemiology, 176*, 512–518. https://doi.org/10.1093/aje/kws128

Benes, F. (2000). Emerging principles of altered neural circuitry in schizophrenia. *Brain Research Review, 31*(2–3), 251–289.

Berger, A., et al. (2021). How is the norepinephrine system involved in the antiepileptic effects of vagus nerve stimulation? *Frontiers Neuroscience, 15*. https://doi.org/10.3389/fnins.2021.790943

Berggren, N., & Derakshan, N. (2013). Attentional control deficits in trait anxiety: Why you see them and why you don't. *Biological Psychology, 92*(3), 440–446. https://doi.org/10.1016/j.biopsycho.2012.03.007

Beucke, J., Sepulcre, J. E., Sebold, M., Kathmann, N., & Kaufmann, C. (2014). Default mode network subsystem alterations in obsessive-compulsive disorder. *British Journal of Psychiatry, 205*, 376–382. https://doi.org/10.1192/bjp.bp.113.137380

Biederman, J. C. (2004). Impact of executive function deficits and attention-deficit/ hyperactivity disorder (ADHD) on academic outcomes in children. *Journal of Clinical and Consulting Psychology, 72(5*, 757–776.

Biederman, J., Monuteaux, M., Doyle, A., Seidman, L., Wilens, T., Ferrero, F., et al. (2004). Impact of executive function deficits and attention-deficit/hyperactivity disorder (ADHD) on academic outcomes in children. *Journal of Consulting and Clinical Psychology, 72*(5), 757–766. https://doi.org/10.1037/0022-006X.72.5.757

Bitler, M., Gelbach, J., & Hoynes, H. (2006). What mean impacts miss: Distributional effects of welfare reform experiments. *American Economic Review, 96*(4), 998–1012. https://doi.org/10.1257/aer.96.4.988

Blum, K., Li-Chuan Chen, A., Braverman, E., Comings, D., Chen, T., & Arcuri, V. (2008). Attention-deficit-hyperactivity disorder and reward deficiency syndrome. *Neuropsychiatric Disease and Treatment, 4*, 5. https://doi.org/10.2147/ndt.s2627

Bogacz, R., Hu, P., & Cohen, J. (2010). Do humans produce the speed–accuracy trade-off that maximizes reward rate? *Quarterly Journal of Experimental Psychology, 63*, 5. https://doi.org/10.1080/17470210903091643

Bora, E. (2020). Meta-analysis of neurocognitive deficits in unaffected relatives of obsessive-compulsive disorder (OCD): Comparison with healthy controls and patients with OCD. *Psychological Medicine, 50*(8), 1257–1266. https://doi.org/10.1017/S0033291720001634

Borji, A. (2012). Boosting bottom-up and top-down visual features for saliency estimation. In *IEEE conference on computer vision and pattern recognition* (pp. 438–445). IEEE.

Bossong, M., & Kahn, R. (2016). The salience of reward. *JAMA Psychiatry, 73*(8), 777–778. https://doi.org/10.1001/jamapsychiatry.2016.1134

Braver, T. (2012). The variable nature of cognitive control: A dual mechanisms framework. *Trends in Cognitive Sciences, 16*(2), 106–113. https://doi.org/10.1016/j.tics.2011.12.010

Brazelton, T. B. (1973). *Neonatal behavioral assessment scale*. Spastics International Medical Publications: Heinemann Medical.

Brieant, A., King-Casas, B., & Kim-Spoon, J. (2022). Transactional relations between developmental trajectories of executive functioning and internalizing and externalizing symptomatology in adolescence. *Developmental Psychopathology, 34*(1), 213–224. https://doi.org/10.1017/S0954579420001054

Broadbent, D. (1958). *Perception and communication*. Oxford University Press.

Bronco, P. (2020). Mapping language with resting-state functional magnetic resonance imaging: A study on the functional profile of the language network. *Human Brain Mapping, 41*(2), 545–560. https://doi.org/10.1002/hbm.24821

Bunford, N., Hamori, G., Nemoda, Z., Angyal, N., Fiath, R. U., & Paszthy, B. (2023). The domain-variant indirect association between electrophysiological response to reward and ADHD presentations is moderated by dopaminergic polymorphisms. *Comprehensive Psychiatry, 124*, 152389. https://doi.org/10.1016/j.comppsych.2023.152389

Bush, G. (2010). Attention-deficit/hyperactivity disorder and attention networks. *Neuropsychopharmacology, 35*, 278–300. https://doi.org/10.1038/npp.2009.120

Cabral, L., Zubiaurre, L., Wild, C., Linke, A., & Cusack, R. (2019). Category-selective visual regions have distinctive signatures of connectivity in neonates. *BioRxiv, 675421*. https://doi.org/10.1101/675421

Carrasco, M. (2011). Visual attention: The past 25 years. *Vision Research, 51*(13), 1484–1525.

Chamberlain, S., Solly, J., Hook, R., Vaghi, M., & Robbin, T. (2021). Cognitive inflexibility in OCD and related disorders. In N. Fineberg & T. Robbins (Eds.), *The neurobiology and treatment of OCD: Accelerating progress* (Current Topics in Behavioral Neurosciences) (Vol. 49). Springer, Switzerland AG. https://doi.org/10.1007/7854_2020_198

Chun, M., & Johnson, M. (2011). Memory: Enduring traces of perceptual and reflective attention. *Neuron, 72*, 520–535. https://doi.org/10.1016/j.neuron.2011.10.026

Churchland, M. M., & Shenoy, K. V. (2007). Temporal complexity and heterogeneity of single-neuron activity in premotor and motor cortex. *Journal of Neurophysiology, 97*, 4235–4257.

Cieslik, E., Mueller, V., Eickhoff, C., Langner, R., & Eickhoff, S. (2015). Three key regions for supervisory attentional control: Evidence from neuroimaging meta-analyses. *Neuroscience & Biobehavioral Reviews, 48*, 22–34.

Clark, L. C.-F. (2017, December 6). Three approaches to understanding and classifying mental disorder: ICD-11, DSM-5, and the National Institute of Mental Health's Research Domain Criteria (RDoC). Retrieved from *Sage Journals Psychological Science in the Public Interest*. https://doi.org/10.1177/1529100617727266

Clark, L., Cuthbert, B., Lewis-Fernandez, R., Narrow, W., & Reed, G. (2017). Three approaches to understanding and classifying mental disorder: ICD-11, DSM-5, and the National Institute of Mental Health's Research Domain Criteria (RDoC). *Psychological Science in the Public Interest, 18*(2). https://doi.org/10.1177/15291006177272

Coghill, D., & Banaschewski, T. (2009). The genetics of attention-deficit/hyperactivity disorder. *Expert Review of Neurotherapeutics, 9*(10), 1547–1565. https://doi.org/10.1586/ern.09.78

Cohen, R. A., Malloy, P. J. M., & Paul, R. (2014). Disorders of attention. In M. Parsons, T. Hammeke, & P. J. Snyder (Eds.), *Clinical neuropsychology: A pocket handbook for assessment* (pp. 463–497). American Psychological Association. https://doi.org/10.1037/14339-021

Connors, K. (2022). *Parent rating scale. (Conners 4™) Conners, Fourth Edition*. WPS Publishing.

Conway, C., & Krueger, R. (2020). Rethinking mental disorder diagnosis: Data-driven psychological dimensions, not categories, as a framework for mental health research, treatment, and training. *Current Directions in Psychological Science*. https://doi.org/10.31234/osf.io/9rx6f

Corbetta, M., & Shulman, G. (2002). Control of goal-directed and stimulus-driven attention in the brain. *Nature Reviews Nueuroscience, 3*, 201–215.

Corlett, P., Honey, G., & Fletcher, P. (n.d.). Prediction error, ketamine and psychosis: An updated model. *Journal of Psychopharmacology, 30*, 1145–1155. https://doi.org/10.1177/0269881116650087

Costa-Dias, T., Wilson, V., Bathula, D., Iyer, S., Mills, K., Thurlow, B., & Stevens, C. (2013). Reward circuit connectivity relates to delay discounting in children with attention-deficit/hyperactivity disorder. *European Neuropsychopharmacology, 23*(1), 33–45. https://doi.org/10.1016/j.euroneuro.2012.10.015

Coull, J., Sahakian, B., & Hodges, J. (1996). The alpha (2) antagonist idazoxan remediates certain attentional and executive dysfunction in patients with dementia of frontal type. *Psychopharmacology (Berlin), 123*(3), 239–249.

Cuthburt, B. (2022). Research domain criteria (RDoC): Progress and potential. *Current Directions of Psychological Science, 31*(2), 107–114. https://doi.org/10.1177/09637214211051363

de la Pena, M., Pan, M., Thai, C., & Alisso, T. (2020). Attention-deficit/hyperactivity disorder predominantly inattentive subtype/presentation: Research progress and translational studies. *Brain Science, 10*(5), 292. https://doi.org/10.3390/brainsci10050292

Derekshan, N., & Eysenck, M. (2009). Anxiety, processing efficiency, and cognitive performance: New developments from attentional control theory. *European Psychologist, 14*(2), 168–176. https://doi.org/10.1027/1016-9040.14.2.168

Derryberry, D., & Reed, M. (2002). Anxiety-related attentional biases and their regulation by attentional control. *Journal of Abnormal Psychology, 111*(2), 225–236. https://doi.org/10.1037/0021-843X.111.2.225

Deserno, M., Bathelt, J., Groenman, A., & Geurts, H. M. (2023). Probing the overarching continuum theory: Data-driven phenotypic clustering of children with ASD or ADHD. *European Child Adolescent Psychiatry, 32*, 1909–1923. https://doi.org/10.1007/s00787-022-01986-9

Diagnostic and statistical manual of mental disorders (5th ed.). (2013). American Psychiatric Association.

Dickstein, S., Bannon, K., Castellanos, F., & Milha, M. (2006). The neural correlates of attention deficit hyperactivity disorder: An ALE meta-analysis. *Journal of Child Psychology and Psychiatry, 47*, 1051–1062. https://doi.org/10.1111/j.1469-7610.2006.01671.x

Dong, G., Lin, X., & Potenza, N. (2015). Decreased functional connectivity in an executive control network is related to impaired executive function in internet gaming disorder. *Progress in Neuro-Psychopharmacology and Biological Psychiatry, 57*, 76–85. https://doi.org/10.1016/j.pnpbp.2014.10.012

Dosenbach, N., Fair, D., Cohen, A. L., Schlaggar, B., & Petersen, S. (2008). A dual-networks architecture of top-down control. *Trends in Cognitive Sciences, 12*(3), 99–105. https://doi.org/10.1016/j.tics.2008.01.001

Drigas, A., & Karyotaki, M. (2019). Attention and its role: Theories and models. *International Journal of Emerging Technologies in Learning, 14*(12), 169–182.

Dudeney, J., Sharpe, L., & Hunt, C. (2015). Attentional bias towards threatening stimuli in children with anxiety: A meta-analysis. *Clinical Psychology Review, 40*, 66–75. https://doi.org/10.1016/j.cpr.2015.05.007

Duller, P., Krishnan, V., & Diaz, A. (2021). A systematic review on the intersection of attention-deficit hyperactivity disorder and gaming disorder. *Journal of Psychiatric Research, 133*, 212–222. https://doi.org/10.1016/j.jpsychires.2020.12.026

Duménieu, M., Marquèze-Pouey, B., Russier, M., & Debanne, D. (2021). Mechanisms of plasticity in subcortical visual areas. *Cells, 10*(11), 3162. https://doi.org/10.3390/cells10113162

Durston, S., van Belle, J., & de Zeeuw, P. (2011). Differentiating frontostriatal and fronto-cerebellar circuits in attention-deficit/hyperactivity disorder. *Biological Psychiatry, 69*(12), 1178–1184. https://doi.org/10.1016/j.biopsych.2010.07.037

Ebrahimpour, M., et al. (2019). Ventral-dorsal neural networks: Object detection via selective attention. In *2019 IEEE winter conference on applications of computer vision (WACV)* (pp. 986–994). IEEE. https://doi.org/10.1109/WACV.2019.00110

Edel, M.-A., Enzi, B., Whitthaus, H., Tegenthoff, M., Peters, S., Juckel, G., & Lissek, S. (2013). Differential reward processing in subtypes of adult attention deficit hyperactivity disorder. *Journal of Psychiatric Research, 47*(3), 350–356. https://doi.org/10.1016/j.jpsychires.2012.09.026

Englemann, J., Damaraju, E., Padmala, S., & Pessoa, L. (2009). Combined effects of attention and motivation on visual task performance: Transient and sustained motivational effects. *Frontiers in Human Neuroscience, 3*. https://doi.org/10.3389/neuro.09.004.2009

Esposito, M. (2007). From cognitive to neural models of working memory. *Philosophical Transactions of the Royal Society of London, 362*(1481), 761–772. https://doi.org/10.1098/rstb.2007.2086

Esposito, R., Cieri, F., Chiacchiaretta, P., Cera, N., Lauriola, M., Di Giannantoni, M., et al. (2018). Modifications in resting state functional anticorrelation between default mode network and dorsal attention network: Comparison among young adults, healthy elders and mild cognitive impairment patients. *Brain Imaging and Behavior, 12*(1), 127–141. https://doi.org/10.1007/S11682-017-9686-Y

Estes, B., & Polnick, B. (2012). Examining motivation theory in higher education: An expectancy theory analysis of tenured faculty productivity. *International Journal of Management, Business, and Administration, 15*(1).

Evarts, E. V. (1868). Relation of pyramidal tract activity to force exerted during voluntary movement. *Journal of Neurophysiology, 31*, 14–27.

Everling, S., Tinsley, C., Gaffan, D., & Duncan, J. (2006). Selective representation of task-relevant objects and locations in the monkey prefrontal cortex. *European Journal of Neuroscience, 23*, 2197–2214. https://doi.org/10.1111/j.1460-9568.2006.04736.x

Eysenck, M. (1992). *Anxiety: The cognitive perspective*. Erlbaum.

Eysenck, M. W., & Calvo, M. (1992). Anxiety and performance: The processing efficiency theory. *Cognition and Emotion, 6*, 409–434.

Eysenck, M., & Derakshan, N. (2011). New perspectives in attentional control theory. *Personality and Individual Differences, 50*(7), 955–960. https://doi.org/10.1016/j.paid.2010.08.019

Eysenck, M., Moser, J., Derakshan, N., & Hepsomali, P. A. (2022). A neurocognitive account of attentional control theory: How does trait anxiety affect the brain's attentional networks? *Cognition and Emotion, 37*(2), 220–237. https://doi.org/10.1080/02699931.2022.2159936

Failing, M., & Theeuwes, J. (2018). Selection history: How reward modulates selectivity of visual attention. *Psychonomic Bulletin & Review, 25*, 514–538. https://doi.org/10.3758/s13423-017-1380-y

Fan, F., Liao, L., Lei, T., Zhao, T., Xia, M., Men, W., et al. (2021). Development of the default-mode network during childhood and adolescence: A longitudinal resting-state fMRI study. *NeuroImage, 226*, 1. https://doi.org/10.1016/j.neuroimage.2020.117581

Feczko, E., Miranda-Dominguez, O., Marr, M., Graham, A., Nigg, A., & Faira, D. (2019). The heterogeneity problem: Approaches to identify psychiatric subtypes. *Trends in Cognitive Science, 23*(7), 584–601. https://doi.org/10.1016/j.tics.2019.03.009

Fitzgerald, K., Stern, E., Angstadt, M., Nicholson-Muth, K., Maynor, M. R., et al. (2010). Altered function and connectivity of the medial frontal cortex in pediatric obsessive-compulsive disorder. *Biological Psychiatry, 68*, 1039–1047. https://doi.org/10.1016/j.biopsych.2010.08.018

Fortunato, C., Bennasar-Vázquez, J., Park, J., Chang, J., Miller, L., Dudman, J., et al. (2024). Nonlinear manifolds underlie neural population activity during behaviour. bioRxiv [Preprint]. 2024 Apr 25:2023.07.18.549575. https://doi.org/10.1101/2023.07.18.549575

Fried, E., & Nesse, R. (2015). Depression is not a consistent syndrome: An investigation of unique symptom patterns in the STAR*D study. *Journal of Affective Disorders, 172*, 96–102. https://doi.org/10.1016/j.jad.2014.10.010

Friston, K. (2008). Hierarchical models in the brain. *PLoS Computational Biology, 4*, e1000211.

Friston, K., Adams, R., Perrinet, L., & Breakspear, M. (2012). Perceptions as hypotheses: Saccades as experiments. *Frontiers in Psychology, 3*, 151.

Fuster, J. (1997). Network memory. *Trends in Neuroscience, 20*(10), 451–459. https://doi.org/10.1016/S0166-2236(97)01128-4

Gallego, J., Perich, M., Naufel, S., Ethier, C., Solla, S., & Miller, E. (2018). Cortical population activity within a preserved neural manifold underlies multiple motor behaviors. *Nature Communications, 9*, 4233. https://doi.org/10.1038/s41467-018-06560-z

Gerstner, W., Kistler, W., Naud, R., & Paninsk, L. (2014). Neuronal Populations. In W. Gerstner, W. Kistler, R. Naud, & L. Paninsk (Eds.), *Neuronal dynamics from single neurons to networks and models of cognition* (pp. 291–322). Cambridge University Press.

Gilbert, J., Wusinich, C., & Zarate, C. (2022). A predictive coding framework for understanding major depression. *Frontiers in Human Neuroscience, 16*. https://doi.org/10.3389/fnhum.2022.787495

Gioia, G., Isquith, P., Retzlaff, P., & Espy, K. (2002). Confirmatory factor analysis of the behavior rating inventory of executive function (BRIEF) in a clinical sample. *Child Neuropsychology, 8*(4), 249–257. https://doi.org/10.1076/chin.8.4.2

Girault, J., Veatch, O., & Won, H. (2024). Etiologic heterogeneity, pleiotropy, and polygenicity in behaviorally defined intellectual and developmental disabilities. *Journal of Neurodevelopmental Disorders, 16*, 8. https://doi.org/10.1186/s11689-024-09526-z

Godefroy, O., & Rousseaux, M. (1996). Divided and focused attention in patients with lesion of the prefrontal cortex. *Brain and Cognition, 30*(2), 155–174. https://doi.org/10.1006/brcg.1996.0010

Goldman-Rakic, P. (1996). Regional and cellular actionation of working memory. *Proceedings of the National Academy of Science iUSA, 93*(24), 13473–13480.

Graziano, P., & Garcia, A. (2016). Attention-deficit hyperactivity disorder and children's emotion dysregulation: A meta-analysis. *Clinical Psychology Review, 46*, 106–123. https://doi.org/10.1016/j.cpr.2016.04.011

Grizenko, N., Paci, M., & Joober, R. (2010). Is the inattentive subtype of ADHD different from the combined/hyperactive subtype? *Journal of Attention Disorders, 13*(6), 649–657. https://doi.org/10.1177/1087054709347200

Gross, J. (2013). Emotion regulation: Taking stock and moving forward. *Emotion, 13*, 359–365. https://doi.org/10.1037/a0032135

Groves, N., Kofler, M., Wells, E., Day, T., & Chan, E. (2020). An examination of relations among working memory, ADHD symptoms, emotion regulation. *Journal of Abnormal Child Psychology, 48*, 525–537. https://doi.org/10.1007/s10802-019-00612-8

Harding, J., et al. (2024). A new predictive coding model for a more comprehensive account of delusions. *Personal View, 11*(4), 295–302. https://doi.org/10.1016/S2215-0366(23)00411-X

Harvey, P. (2019). Domains of cognition and their assessment. *Dialouges in Clinical Neuroscience*, 227–237. https://doi.org/10.31887/DCNS.2019.21.3/pharvey

Hermans, E., Henckens, M., & Joëls, M. G. (2014). Dynamic adaptation of large-scale brain networks in response to acute stressors. *Trends in Neuroscience, 37*, 304–314. https://doi.org/10.1016/j.tins.2014.03.006

Hester, R., & Garavan, H. (2005). Working memory and executive function: The influence of content and load on the control of attention. *Memory and Cognition, 33*, 221–233. https://doi.org/10.3758/BF03195311

Hillyard, S., Vogel, E., & Luck, S. (1998). Sensory gain control (amplification) as a mechanism of selective attention: Electrophysiological and neuroimaging evidence. *Philosophical Transactions of the Royal Society of London on Biological Science, 353*(1373), 1257–1270. https://doi.org/10.1098/rstb.1998.0281

Hofman, M. (2014). Evolution of the human brain: When bigger is better. *Frontiers Neuroanatomy, 8*. https://doi.org/10.3389/fnana.2014.00015

Hohwy, J. (2013). *The predictive mind*. Oxford University Press.

Holland, N., Robbins, T., & Rowe, J. (2021). The role of noradrenaline in cognition and cognitive disorders. *Brain, 144*, 2243–2256. https://doi.org/10.1093/brain/awab111

Hovel, H., Viol, A., Loske, P., Merfort, L., & Vuksanović, V. (2020). Synchronization in functional networks of the human brain. *Journal of Non Linear Science, 30*, 2259–2282. https://doi.org/10.1007/s00332-018-9505-7

Hsu, K., Beard, C., Rifkin, L., Dillon, D., Pizzagalli, D., & Björgvinsson, T. (2015). Transdiagnostic mechanisms in depression and anxiety: The role of rumination and attentional control. *Journal of Affective Disorders, 188*, 22–27. https://doi.org/10.1016/j.jad.2015.08.008

Hu, B., Rao, J., Li, X., Cao, T. L., Majoe, D., & Gutknecht, J. (2017). Emotion regulating attentional control abnormalities in major depressive disorder: An event-related potential study. *Scientific Reports, 7*, 13530. https://doi.org/10.1038/s41598-017-13626-3

Humphrey, D. R., & Thompson, W. D. (1970). Predicting measures of motor performance from multiple cortical spike trains. *Science, 170*(7), 758–762.

Hyman, A. (2005). Addiction: A disease of learning and memory. *The American Journal of Psychiatry*. https://doi.org/10.1176/appi.ajp.162.8.1414

Insel, T. (2014). The NIMH research domain criteria (RDoC) project: Precision medicine for psychiatry. *The American Journal of Psychiatry, 171*(4), 395–397. https://doi.org/10.1176/appi.ajp.2014.14020138

Itti, L., & Baldi, P. (2006). Bayesian surprise attracts human attention. *Advances in Neural Information Processing Systems, 18*, 547.

James, H. (1890). *The principles of psychology*. Holt.

Jankowiak-Siuda, K., & Zajkowski, W. (2013). A neural model of mechanisms of empathy deficits in narcissism. *Medical Science Monitor, 19*, 934–941. https://doi.org/10.12659/MSM.889593

Jensen, C., & Steinhausen, H. (2015). Comorbid mental disorders in children and adolescents with attention-deficit/hyperactivity disorder in a large nationwide study. *ADHD Attention Deficit and Hyperactivity Disorders, 7*(1), 27–38.

Joel, D. W. (2000). The connections of the dopaminergic system with the striatum in rats and primates: An analysis with respect to the functional and compartmental organization of the striatum. *Neuroscience*, 451–474.

Johnson, M., Gilga, T., & Charman, T. (2014). Annual research review: Infant development, autism, and ADHD—Early pathways to emerging disorders. *The Journal of Child Psychology and Psychiatry*. https://doi.org/10.1111/jcpp.12328

Kalkbrenner, A., Zheng, C., Yu, J., Jenson, T., Kuhlwein, T., Ladd-Acosta, C., et al. (2024). Method for testing etiologic heterogeneity among noncompeting diagnoses, applied to impact of perinatal exposures on autism and attention deficit hyperactivity disorder. *Epidemiology, 35*(5), 689–700. https://doi.org/10.1097/EDE.0000000000001760

Kanwisher, N., McDermott, J., & Chun, M. (1997). The fusiform face area: A module in human extrastriate cortex specialized for face perception. *Journal of Neuroscience*, 4302–4311.

Kapser, L., Aldrson, R., & Hudec, K. (2012). Moderators of working memory deficits in children with attention-deficit/hyperactivity disorder (ADHD): A meta-analytic review. *Clinical Psychology Review, 32*(7), 605–617.

Karalunas, S., Fair, D., Musser, E., Aykes, K., Iyer, S., & Nigg, J. (2014). Subtyping attention-deficit/hyperactivity disorder using temperament dimensions: Toward biologically based nosologic criteria. *JAMA Psychiatry, 71*, 1015–1024. https://doi.org/10.1001/jamapsychiatry.2014.763

Katsuki, F., & Constantinidis, C. (2014). Bottom-up and top-down attention: Different processes and overlapping neural systems. *The Neuroscientist, 20*, 509–521. https://doi.org/10.1177/1073858413514136

Katsuki, F., & Constantinidis, C. (2016). Bottom-up and top-down attention: Different processes and overlapping neural systems. *The Neuroscientist, 201X, Vol XX(X)*, 1–13. https://doi.org/10.1177/1073858413514136

Keller, A., et al. (2019). Paying attention to attention in depression. *Translational Psychiatry, 9*, 279. https://doi.org/10.1038/s41398-019-0616-1

Kertzman, S., Reznik, I., Hornik-Lurie, T., Weizman, A., Kotler, M., & Amital, D. (2010). Stroop performance in major depression: Selective attention impairment or psychomotor slowness? *Journal of Affective Disorders, 122*(1–2), 167–173. https://doi.org/10.1016/j.jad.2009.08.009

Kessler, R. C., et al. (2006). The prevalence and correlates of adult ADHD in the United States: Results from the National Comorbidity Survey Replication. *American Journal of Psychiatry, 163*(4), 716–723.

Kim, N., & Kastner, S. (2019). A biased competition theory for the developmental cognitive neuroscience of visuo-spatial attention. *Current Opinion in Psychology, 29*, 219–228. https://doi.org/10.1016/j.copsyc.2019.03.017

Kimura, A., Yonetani, R., & Hirayami, T. (2013). Computational models of human visual attention and their implementations: A survey. *IEICE Transactions on Information and Systems, E96-D*(3), 562–578, 562–578.

Knill, D., & Pouget, A. (2004). The Bayesian brain: The role of uncertainty in neural coding and computation. *Trends in Neuroscience, 27*, 712–719. https://doi.org/10.1016/j.tins.2004.10.007

Koch, C., & Ullman, S. (1985). Shifts in selective visual attention: Towards the underlying neural circuitry. *Human Neurobiology, 4*(2), 19–27.

Kofler, M., Rapport, M., Boldin, J., Sarver, D., & Raiker, J. (2009). ADHD and working memory: The impact of central executive deficits and exceeding storage/rehearsal capacity on observed inattentive behavior. *Journal of Abnormal Child Psychology, 38*, 149–161. https://doi.org/10.1007/s10802-009-9357-6

Kolb, B. M. (2012, October 16). Experience and the development of the prefrontal cortex. *Proceedings of the National Academy of Science, 109*.

Kowalski, J., Wypych, M., Marchewka, A., & Dragan, M. (2019). Neural correlates of cognitive-attentional syndrome: An fMRI study on repetitive negative thinking induction and resting state functional connectivity. *Frontiers in Psychology, 10*. https://doi.org/10.3389/fpsyg.2019.00648

Koziol, L. F., & Budding, D. E. (2009). *Subcortical structures and cognition: Implications for neuropsychological assessment*. Springer.

Koziol, L., & Lutz, J. (2013). From movement to thought: The development of executive function. *Applied Neuropsychology: Child, 2*(2), 104–115. https://doi.org/10.1080/21622965.2013.748386

Koziol, A., Joyce, A., & Wurglitz, G. (2014). The neuropsychology of attention: Revisiting the "Mirsky model". *Applied Neuropsychology: Child, 3*(4). https://doi.org/10.1080/21622965.2013.870016

Kunsti, J., et al. (2014). The separation of ADHD inattention and hyperactivity-impulsivity symptoms: Pathways from genetic effects to cognitive impairments and symptoms. *Journal of Abnormal Child Psychology*, 127–136. https://doi.org/10.1007/s10802-013-9771-7

Kuppens, P. (2006). Individual differences in the relationship between pleasure and arousal. *Journal of Research in Personality, 42*(4), 1053–1059. https://doi.org/10.1016/j.jrp.2007.10.007

Le-Cunff, A. (2022). *Is there a limit to the human brain's capacity?* Retrieved from Ness labs: https://nesslabs.com/brain-limitations

Lee, D., Quattrocki Knight, E., Song, H., Lee, S., Pae, C., Yoo, S., & Park, H. (2021). Differential structure-function network coupling in the inattentive and combined types of attention deficit hyperactivity disorder. *PLoS One, 16*(12), e0260295. https://doi.org/10.1371/journal.pone.0260295

Legrain, V., Plaghki, L., & Mouraux, A. (2011). The pain matrix reloaded. *Progress in Neurobiology, 93*, 111–124. https://doi.org/10.1016/j.pneurobio.2010.10.005

Levy, N. (2018). Obsessive–compulsive disorder as a disorder of attention. *Mind and Language, 33*(1), 3–16. https://doi.org/10.1111/mila.12172

Lewis-Peacock, J., Kessler, Y., & Oberauer, K. (2018). The removal of information from working memory. *Annals of the New York Academy of Sciences, 1424*(1), 33–44. https://doi.org/10.1111/nyas.13714

Liebel, S., & Nelson, J. (2017). Auditory and visual working memory functioning in college students with attention-deficit/hyperactivity disorder and/or learning disabilities. *Archives of Clinical Neuropsychology, 32*(8), 980–991. https://doi.org/10.1093/arclin/acx014

Limongi, R., et al. (2020). Glutamate and dysconnection in the salience network: Neurochemical, effective connectivity and computational evidence in schizophrenia. *Biological Psychiatry, 88*, 273–281. https://doi.org/10.1016/j.biopsych.2020.01.021

Lintz, E., & Johnson, M. (2021). Refreshing and removing items in working memory: Different approaches to equivalent processes? *Cognition, 211*, 104655. https://doi.org/10.1016/j.cognition.2021

Liu, J. C., Liu, K., & Zhao, J. (2018). Event detection via gated multilingual attention mechanism. In *Thirty-second AAAI conference on artificial intelligence*. Association for the Advancement of Artificial Intelligence.

Lubke, G. H., Derkes, E., van Bijsterveldt, T., & Boomsma, D. (2009). Maternal ratings of attention problems in ADHD: Evidence for the existence of a continuum. *Journal of the American Academy of Child & Adolescent Psychiatry, 48*(11), 1085–1093. https://doi.org/10.1097/CHI.0b013e3181ba3dbb

Luo, Y., Weibman, D., Halperin, J., & Li, X. (2019). A review of heterogeneity in attention deficit/hyperactivity disorder (ADHD). *Frontiers in Human Neuroscience, 13*, 42. https://doi.org/10.3389/fnhum.2019.00042

Lux, V. (2018). Epigenetic programming effects of early life stress: A dual-activation hypothesis. *Current Genomics., 19*(8), 638–652. https://doi.org/10.2174/1389202919666180307151358

Mackintosh, N. J. (2016). A theory of attention: Variations in the associability of stimuli with reinforcement. *Psychological Review, 82*(4), 276–298. https://doi.org/10.1037/h0076778

MacLeod, C., & Clarke, P. (2015). The attentional bias modification approach to anxiety intervention. *Clinical Psychological Science, 3*(1), 58–78. https://doi.org/10.1177/2167702614560749

MacLeod, C., Rutherford, E., Campbell, L., Ebsworthy, G., & Holker, L. (2002). Selective attention and emotional vulnerability: Assessing the causal basis of their association through the experimental manipulation of attentional bias. *Journal of Abnormal Psychology, 11*, 107–123.

Maehler, C., & Schuchardt, K. (2016). Working memory in children with specific learning disorders and/or attention deficits. *Learning and Individual Differences, 49*, 341–347. https://doi.org/10.1016/j.lindif.2016.05.007

Magnus, W. N. (2023, August 8). *Attention deficit hyperactivity disorder*. Retrieved from National Library of Medicine National Center for Biotechnology Information: https://www.ncbi.nlm.nih.gov/books/NBK441838/

Maiese, M. (2023). Are all mental disorders affective disorders? *Passion: Journal of the European Philosophical Society for the Study of Emotions, 1*(1), 31–49. https://doi.org/10.59123/passion.v1i1.12728

Makovac, E., Fagioli, S., Rae, C. L., Critchley, H., & Ottaviani, C. (2020). Can't get it off my brain: Meta-analysis of neuroimaging studies on perseverative cognition. *Psychiatry Research: Neuroimaging, 295*, 111020. https://doi.org/10.1016/j.pscychresns.2019.111020

Manoliu, A., et al. (2014). Insular dysfunction within the salience network is associated with severity of symptoms and aberrant inter-network connectivity in major depressive disorder. *Frontiers in Human Neuroscience*. https://doi.org/10.3389/fnhum.2013.00930

Mao, A., & Findling, R. (2014). Comorbidities in adult attention-deficit/hyperactivity disorder: A practical guide to diagnosis in primary care. *Postgraduate Medicine, 126*(5), 42–51. https://doi.org/10.3810/pgm.2014.09.2799

Mariotti, P., Mariotti, P., Lucibello, S., De Clemente, V., Veltri, S., Janiri, D., et al. (2021). ADHD syndrome and tic disorders: Two sides of the same coin. *The Lancet*. https://doi.org/10.2139/ssrn/3923340

Markett, S., Nothdurfter, D., Fosca, A., & Reuter, M. (2022). Attention networks and the intrinsic network structure of the human brain. *Human Brain Mapping, 43*, 1431–1448. https://doi.org/10.1002/hbm.25734

Martinussen, R., Hayden, J. H.-J., & Tannock, R. (2005). A meta-analysis of working memory impairments in children with attention-deficit/hyperactivity disorder. *Journal of the American Academy of Child & Adolescent Psychiatry*, 377–384. https://doi.org/10.1097/01.chi.0000153228.72591.73

Mattfield, A., Whitfield-Gabrieli, S., Biederman, J., Spencer, T., Brown, A., et al. (2016). Dissociation of working memory impairments and attention-deficit/hyperactivity disorder in the brain. *NeuroImage: Clinical, 10*, 274–282. https://doi.org/10.1016/j.nicl.2015.12.003

Mayes, S., Calhoun, S., Mayes, R., & Molitoris, S. (2012). Autism and ADHD: Overlapping and discriminating symptoms. *Research in Autism Spectrum Disorders, 6*(1), 277–285. https://doi.org/10.1016/j.rasd.2011.05.009

McCormick, E., Arnemann, K., Ito, T., Hanson, S., & Cole, M. (2022). Latent functional connectivity underlying multiple brain states. *Network Neuroscience, 6*(2), 570–590. https://doi.org/10.1162/netn_a_00234

McLennan, J. (2016). Understanding attention deficit hyperactivity disorder as a continuum. *The Official Journal of the College of Family Physicians Canada, 62*(12), 979–982.

McNicol, D. (2004). *A primer of signal detection theory*. Psychology Press. https://doi.org/10.4324/9781410611949

Meacham, F., & Bergstrom, C. (2016). Adaptive behavior can produce maladaptive anxiety due to individual differences in experience. *Evolution, Medicine, and Public Health, 2016*(1), 270–285. https://doi.org/10.1093/emph/eow02

Meek, W., & Benson, A. (2002). Dissecting the brain's internal clock: How frontal–striatal circuitry keeps time and shifts attention. *Brain and Cognition, 48*(1), 195–211. https://doi.org/10.1006/brcg.2001.1313

Melloni, L., van Leeuwen, S., Alink, A., & Muller, N. (2012). Interaction between bottom-up saliency and top-down control: How saliency maps are created in the human brain. *Cerebral Cortex, 22*(12), 2943–2952. https://doi.org/10.1093/cercor/bhr384

Menon, V. (2015). *Brain mapping: An encyclopedic reference* (Vol. 2). Academic press.

Menon, V., & Uddin, L. (2010). Saliency, switching, attention and control: A network model of insula function. *Brain Structure and Function, 214*(5-6), 655–667. https://doi.org/10.1007/s00429-010-0262-0

Meyer, K., & Damasio, A. (2009). Convergence and divergence in a architecture for recognition and memory. *Trends in Neuroscience, 32*(7), 376–382. https://doi.org/10.1016/j.tins.2009.04.002

Milich, R. B., & Lynam, D. (2001). ADHD combined type and ADHD predominantly inattentive type are distinct and unrelated disorders. *Clinical Psychology: Science and Practice, 8*(4), 463–488. https://doi.org/10.1093/clipsy.8.4.463

Mirsky, A. F., et al. (1991). Analysis of the elements of attention: A neuropsychological approach. *Neuropsychology Review, 2*, 109–145.

Mitchell-Hegg, R., Prado, S., Gava, G., Go, M., & Schultz, S. (2023). Neural manifold analysis of brain circuit dynamics in health and disease. *Journal of Consultational Neuroscience, 51*(1), 1–21. https://doi.org/10.1007/s10827-022-00839-3

Mohr, K., Carr, N., Georgel, R., & Kelly, S. (2020). Modulation of the earliest component of the human VEP by spatial attention: An investigation of task demands. *Cerebral Cortex Communications, 1*(1), tgaa045.

Molloy, M. S. (2022). Individual variability in functional organization of the neonatal brain. *NeuroImage, 253*, 119101. https://doi.org/10.1016/j.neuroimage.2022.119101

Monk, C., et al. (2008). Amygdala and nucleus accumbens activation to emotional facial expressions in children and adolescents at risk for major depression. *American Journal of Psychiatry, 165*(1), 90–98.

Mulder, H., Pitchford, N., Hagger, M., & Marlow, N. (2009). Development of executive function and attention in preterm children: A systematic review. *Developmental Neuropsychology, 34*(4), 393–421. https://doi.org/10.1080/87565640902964524

Muller, J., & Roberts, J. (2005). Memory and attention in obsessive–compulsive disorder: A review. *Journal of Anxiety Disorders, 19*(1), 1–26. https://doi.org/10.1016/j.janxdis.2003.12.001

Murty, V., & Adcock, R. (2014). Enriched encoding: Reward motivation organizes cortical networks for hippocampal detection of unexpected events. *Cerebral Cortex, 24*(8), 2160–2168.

Najt, P., Glahn, D., Bearden, C., Hatch, J. P., et al. (2005). Attention deficits in bipolar disorder: A comparison based on the Continuous Performance Test. *Neuroscience Letter, 379*, 122–126. https://doi.org/10.1016/j.neulet.2004.12.051

Nazari, R., & Salehi, M. (2023). Early development of the functional brain network in newborns. *Brain Structure & Function, 228*, 1725–1739. https://doi.org/10.1007/s00429-023-02681-4

Nelson, A., & Bender, K. (2021). Dendritic integration dysfunction in neurodevelopmental disorders. *Developmental Neuroscience, 43*(3–4), 201–221. https://doi.org/10.1159/000516657

Newell, A. (1956). The logic theory machine: A complex information processing system. *IRE Transactions on Information Theory, 2*(3), 6179. https://doi.org/10.1109/TIT.1956.1056797

Nigg, J., & Casey, B. (2005). An integrative theory of attention-deficit/ hyperactivity disorder based on the cognitive and affective neurosciences. *Developmental Psychopathology, 17*, 785–806. https://doi.org/10.1017/s0954579405050376

Oberauer, K. (2019). Working memory and attention—A conceptual analysis and review. *Journal of Cognition, 2*(1), 36. https://doi.org/10.5334/joc.58

Oldham, S., & Fornito, A. (2019). The development of brain network hubs. *Developmental Cognitive Neuroscience, 36*, 100607.

Oldham, S., Fulcher, B., Parkes, L., Arnatkevičiūtė, A., Suo, C., & Fornito, A. (2019). Consistency and differences between centrality measures across distinct classes of networks. *PLoS One, 14*(7), e0220061. https://doi.org/10.1371/journal.pone.0220061

Onofrj, V., Chiarelli, A., Wise, R. C., & Caulo, M. (2022). Interaction of the salience network, ventral attention network, dorsal attention network and default mode network in neonates and early development of the bottom-up attention system. *Brain Structure and Function*. https://doi.org/10.1007/s00429-022-02477-y

Pace-Schott, E. P. (2017). *In principles and practice of sleep medicine*. Elsevier.

Pardo, J., Pardo, P., & Janer, K. R. (1990). The anterior congulate cortex, mediates processing selection in the Stroop attentional conflict paradigm. *PNAS, 86*(1), 256–259.

Park, H., & Friston, K. (2013). Structural and functional brain networks: From connections to cognition. *Science, 342*(6158), 1238411. pmid:24179229.

Parr, T., & Friston, K. (2019). Attention or salience? *Current Opinion in Psychology, 29*, 1–5. https://doi.org/10.1016/j.copsyc.2018.10.006

Paulus, F., Ohmann, S., Plenar, P., & Popow, C. (2021). Emotional dysregulation in children and adolescents with psychiatric disorders. A narrative review. *Frontiers in Psychiatry, 12*. https://doi.org/10.3389/fpsyt.2021.628252

Pavlov, I. (1927). *Conditioned reflexes: An investigation of the physiological activity of the cerebral cortex*. Oxford University Press.

Peelan, M., Heslenfeld, D., & Theeuwes, J. (2004). Endogenous and exogenous attention shifts are mediated by the same large-scale neural network. *NeuroImage, 22*(2), 822–830. https://doi.org/10.1016/j.neuroimage.2004.01.044

Penrose, O., & liebowitz, J. (1979). Toward a rigorous molecular theory of Metastability. In E. Montroll & J. Leibowitz (Eds.), *Studies in statisical mechanics II, fluctuation phenomena* (pp. 319–322). North Holland Publishing Company.

Petersen, S., & Posner, M. (2012). The attention system of the human brain: 20 years after. *Annual Review of Neuroscience, 2012*(35), 73–89. https://doi.org/10.1146/annurev-neuro-062111-150525. PMID: 22524787; PMCID: PMC3413263.

Petersen, S., Fox, P., Miezin, F., & Raichle, M. (1988). Modulation of cortical visual responses by direction of spatial attention measured by PET. *Association for Research in Vision Ophthalmology, 22*.

Pfiffner, L., Mikami, A., Huang-Pollack, C., Easterlin, B., Zalecki, C., & McBurnett, K. (2007). A randomized, controlled trial of integrated home-school behavioral treatment for ADHD, predominantly inattentive type. *Journal of the American Academy of Child & Adolescent Psychiatry, 46*(8), 1041–1050. https://doi.org/10.1097/chi.0b013e318064675f

Pharisi, D., & Nolfi, F. (2001). *Development in Neural Networks*. Retrieved from MIT Press Direct: https://direct.mit.edu/books/edited-volume/2039/chapter-abstract/55299/Development-in-Neural-Network

Phillips, M., Ladouceur, C., & Drevets, W. (2008). A neural model of voluntary and automatic emotion regulation: Implications for understanding the pathophysiology and neurodevelopment of bipolar disorder. *Molecular Psychiatry, 13*, 829, 833–857.

Piani, M., Maggioni, E., Delvecchio, G., & Brambilla, P. (2022). Sustained attention alterations in major depressive disorder: A review of fMRI studies employing Go/no-Go and CPT tasks. *Journal of Affective Disorders, 303*, 98–113. https://doi.org/10.1016/j.jad.2022.02.003

Pinto, Y., van der Leij, A., Sligte, S., Lamme, V., & Scholte, S. (2013). Bottom-up and top-down attention are independent. *Journal of Vision, 13*, 16. https://doi.org/10.1167/13.3.16

Posner, M., & Petersen, S. (1990). The attention system of the human brain. *Annual Review of Neuroscience, 13*(1), 25–42. https://doi.org/10.1146/annurev.ne.13.030190.000325

Posner, J., Song, I., Lee, S., Rodriguez, C. I., Moore, H., Marsh, R., & Blair Simpson, H. (2017). Increased functional connectivity between the default mode and salience networks in unmedicated adults with obsessive-compulsive disorder. *Human Brain Mapping, 38*(2), 678–687.

Posner, J., Polanczyk, G., & Sonuga-Barke, E. (2020). Attention-deficit hyperactivity disorder. *Lancet, 395*(10), 450–462. https://doi.org/10.1016/S0140-6736(19)33004-1

Raichle, M. (2015). The brain's default mode network. *Annual Review of Neuroscience, 38*(July), 433–447. https://doi.org/10.1146/ANNUREV-NEURO-071013-014030

Raiker, J. S., Friedman, L. M., Orban, S., Kofler, M. J., Sarver, D. E., & Rapport, M. D. (2019). Phonological working memory deficits in ADHD revisited: The role of lower level information-processing deficits in impaired working memory performance. *Journal of Attention Disorders, 23*(6), 570–583. https://doi.org/10.1177/1087054716686182

Rapport, M., Bolden, J., Kofler, M., Sarver, D., Raiker, J., & Alderson, R. (2009). Hyperactivity in boys with attention-deficit/hyperactivity disorder (ADHD): A ubiquitous core symptom or manifestation of working memory deficits? *Journal of Abnormal Child Psychology, 37*, 521–534.

Raye, C., Johnson, M., Mitchell, E., Greene, K., et al. (2007). Refreshing: A minimal executive function. *Cortex, 43*, 145. https://doi.org/10.1016/S0010-9452(08)70451-9

Reale, L., Bartoli, B., Cartabia, M., Zanetti, M., Costantino, M., et al. (2017). Comorbidity prevalence and treatment outcome in children and adolescents with ADHD. *European Child and Adolescent Psychiatry, 26*, 1443–1457. https://doi.org/10.1007/s00787-017

Reimherr, F., Roesler, M., Marchant, B., Gift, T., Tetz, W., & Philipp-Wiegmann, F. (2020). Types of adult attention-deficit/hyperactivity disorder: A replication analysis. *The Journal of Clinical Psychiatry, 8*, 12. https://doi.org/10.4088/JCP.19m13077

Richer, F., Decary, A., Lapierre, M., Rouleau, I., Bouvier, G., & Sainthilaire, J. (1993). Target detection deficits in frontal lobectomy. *Brain and Cognition, 21*(2), 203–211. https://doi.org/10.1006/brcg.1993.1016

Roberts, C. S. (2023). *Impact and centrality of attention dysregulation on cognition, anxiety and low mood in adolescents*. Scientific Reports.

Roda, C., & Thomas, J. (2006). Attention aware systems: Theories ,applications, and research agenda. *Computers in Human Behavior, 22*, 557–587.

Rolls, E., & Stringer, S. (2000). On the design of neural networks in the brain by genetic evolution. *Progress in Neurobiology, 61*(6), 557–579. https://doi.org/10.1016/S0301-0082(99)00066-0

Rolls, E., McCabe, C., & Redoute, J. (2008). Expected value, reward outcome, and temporal difference error representations in a probabilistic decision task. *Cerebral Cortex, 18*(3), 652–663.

Romer, A., & Pizzagalli, D. (2021). Is executive dysfunction a risk marker or consequence of psychopathology? A test of executive function as a prospective predictor and outcome of general psychopathology in the adolescent brain cognitive development study. *Developmental Cognitive Neuroscience., 51*, 100994. https://doi.org/10.1016/j.dcn.2021.100994

Rosales, A., Vitoratou, S., Banaschewski, T., Asherson, P., Buitelaar, J., Oades, R., et al. (2015). Are all the 18 DSM-IV and DSM-5 criteria equally useful for diagnosing ADHD and predicting comorbid conduct problems? *European Child Adolescent Psychiatry, 24*, 1325–1337. https://doi.org/10.1007/s00787-015-0683-7

Ross, A., & Thompson, R. (2011). Emotion and emotion regulation: Two sides of the developing coin. *Emotion Review, 3*, 53–61. https://doi.org/10.1177/1754073910380969

Rost, T., Deger, M., & Nawrot, M. (2018). Winnerless competition in clustered balanced networks: Inhibitory assemblies do the trick. *Biological Cybernetics, 112*(5283). https://doi.org/10.1007/s00422-017-0737-7

Saad, J., et al. (2022). Intrinsic functional connectivity in the default mode network differentiates the combined and inattentive attention deficit hyperactivity disorder types. *Frontiers in Human Neuroscience, 09*, 16. https://doi.org/10.3389/fnhum.2022.859538

Saboory, E., Ghasemi, M., & Mehranfard, D. (2020). Norepinephrine, neurodevelopment and behavior. *Neurochemistry International, 135*, 104706. https://doi.org/10.1016/j.neuint.2020.104706

Salehinejad, M., Ghayerin, E., Nejati, V., & Yavari, F. N. (2020). Domain-specific involvement of the right posterior parietal cortex in attention network and attentional control of ADHD: A randomized, cross-over, sham-controlled tDCS study. *Neuroscience, 444*, 149–159. https://doi.org/10.1016/j.neuroscience.2020.07.037

Sali, A., Aderson, B., & Yantis, S. (2014). The role of reward prediction in the control of attention. *Journal of Experimental Psychology: Human Perception and Performance, 40*(4), 1654–1664. https://doi.org/10.1037/a0037267

Salmi, S., Rinne, T., Koistien, S., Salonen, O., & Alho, K. (2009). Brain networks of bottom-up triggered and top-down controlled shifting of auditory attention. *Brain Research, 1286*, 155–164. https://doi.org/10.1016/j.brainres.2009.06.083

Salmi, J., Salmela, V., Salmo, E., Mikkola, K., Leppämäki, S., Tani, P., & Hokkanen, H. (2018). Out of focus—Brain attention control deficits in adult ADHD. *Brain Research, 1692*, 12–22. https://doi.org/10.1016/j.brainres.2018.04.019

Salum, G., Gadelha, A., Polanczyk, G. M., & Rohde, L. (2018). Diagnostic operationalization and phenomenological heterogeneity in psychiatry: The case of attention deficit hyperactivity disorder. *Salud Mental, 41*(6) México Nov./Dec. https://doi.org/10.17711/sm.0185-3325.2018.037

Sanchez-Lopez, A., Koster, E., & Van Put, J. (2019). Attentional disengagement from emotional information predicts future depression via changes in ruminative brooding: A five-month longitudinal eye-tracking study. *Behavior Research and Therapy, 118*, 30–42. https://doi.org/10.1016/j.brat.2019.03.013. Epub 2019 Mar 27.

Sandstrom, A., Perroud, N., Alda, M., Uher, R., & Pavlova, B. (2021). Prevalence of attention-deficit/hyperactivity disorder in people with mood disorders: A systematic review and meta-analysis. *Acta Neuropsychiatrica Scandinavica, 143*(5), 380–391. https://doi.org/10.1111/acps.13283

Sani, I., Stemmann, H., Caron, B., & al., e. (2021). The human endogenous attentional control network includes a ventro-temporal cortical node. *Nature Communications, 12*, 360. https://doi.org/10.1038/s41467-020-20583-5

Sanislow, C., Quinn, K. J., & Sypher, I. (2015). NIMH research domain criteria (RDoC). In R. Cautin & S. Lilienfeld (Eds.), *The encyclopedia of clinical psychology*. Wiley-Blackwell.

Sato, T., Watanabe, K., Thompson, K., & Schall, J. (2003). Effect of target-distractor similarity on FEF visual selection in the absence of the target. *Experimental Brain Research, 151*(3), 356–363.

Scahachar, R. (2023). Fifty years of executive control research in attention-deficit/hyperactivity disorder: What we have learned and still need to know. *Neuroscience & Biobehavioral Reviews, 155*, 105461. https://doi.org/10.1016/j.neubiorev.2023.105461

Schimmelpfennig, J., Topczewski, J., Zajkowski, W., & Jankowiak-Siuda, K. (2023). The role of the salience network in cognitive and affective deficits. *Frontiers in Human Neuroscience, 17*, 1133367. https://doi.org/10.3389/fnhum.2023.1133367

Schneidt, S., Jusyte, A., Rauss, K., & Schönenberg, M. (2018). Distraction by salient stimuli in adults with attention-deficit/hyperactivity disorder: Evidence for the role of task difficulty in bottom-up and top-down processing. *Cortex, 101*, 206–220. https://doi.org/10.1016/j.cortex.2018.01.021

Schweitzer, J., Hanford, R., & Medoff, D. (2006). Working memory deficits in adults with ADHD: Is there evidence for subtype differences? *Brain and behavioral Functions, 2*, 43. https://doi.org/10.1186/1744-9081-2-43

Science Direct. (2020). *Precuneus*. Retrieved from Science Direct: https://www.sciencedirect.com/topics/neuroscience/precuneus#:~:text=The%20precuneus%20is%20a%20brain,and%20affective%20responses%20to%20pain

Scott, S. H., & Kalaska, J. F. (1997). Reaching movements with similar hand paths but different arm orientations. I. Activity of individual cells in motor cortex. *Journal of Neurophysiology*, 826–852.

Seeley, W. (2019). The salience network: A neural system for perceiving and responding to homeostatic demands. *Journal of Neuroscience, 39*(50), 9878–9882. https://doi.org/10.1523/JNEUROSCI.1138-17.2019

Seth, A., Suzuki, K., & Critchley, H. (2011). An interoceptive predictive coding model of conscious presence. *Frontiers in Psychology, 2*, 395. https://doi.org/10.3389/fpsyg.2011.00395

Shaw, P., Stringaris, A., Nigg, J., & Leibenluft, E. (2014). Emotion dysregulation in attention deficit hyperactivity disorder. *The American Journal of Psychiatry*. https://doi.org/10.1176/appi.ajp.2013.13070966

Shen, K., et al. (2012). Information processing architecture of functionally defined clusters in the macaque cortex. *Journal of Neuroscience, 32*(48), 17465–17476. Epub 2012/12/01. pmid:23197737.

Shiels, K., & Hawk, L., Jr. (2010). Self-regulation in ADHD: The role of error processing. *Clinical Psychology Review, 30*(8), 951–961. https://doi.org/10.1016/j.cpr.2010.06.010

Singer, W. (1993). Synchronization of cortical activity and its putative. *Annual Review of Physiology, 55*, 349–374.

Sinnott-Armstrong, W., Soter, L., & Summers, J. (2024). Mental disorders as failures of attention. *Crítica. Revista Hispanoamericana De Filosofía*, 17–44. https://doi.org/10.22201/iifs.18704905e.2024.1528

Skirrow, C., & Asherson, P. (2013). Emotional lability, comorbidity and impairment in adults with attention-deficit hyperactivity disorder. *Journal of Affective Disorders, 147*(1-3), 273–281.

Sohn, H., Meirhaeghe, N., Rajalingham, R., & Jazayeri, M. (2021). A network perspective on sensorimotor learning. *Trends in Neuroscience, 44*(3), 170–181. https://doi.org/10.1016/j.tins.2020.11.007

Solanto, M., Shuktz, K., Fan, J., Tang, C., & Newcorn, J. (2009). Event-related fMRI of inhibitory control in the predominantly inattentive and combined subtypes of ADHD. *Journal of Neuroimaging*. https://doi.org/10.1111/j.1552-6569.2008.00289.x

Song, H., Shim, W., & Rosenebrg, M. (2023). Large-scale neural dynamics in a shared low-dimensional state space reflect cognitive and attentional dynamics. *Neuroscience*. https://doi.org/10.7554/eLife.85487

Sonuga-Barke, E. (2005). Causal models of attention-deficit/hyperactivity disorder: From common simple deficits to multiple developmental pathways. *Biological Psychiatry, 57*, 1231–1238.

Stam, C. (2024). Hub overload and failure as a final common pathway in neurological brain network disorders. Network. *Neuroscience, 8*(1), 1–23. https://doi.org/10.1162/netn_a_00339

Sternheim, L., van Passal, B., Dingermans, A., Cath, D., & Danner, U. (2022). Cognitive and experienced flexibility in patients with anorexia nervosa and obsessive compulsive disorder. *Frontiers Psychiatry, 16*. https://doi.org/10.3389/fpsyt.2022.868921

Sterzer, P., Voss, M., Schlagenhauf, F., & Heinz, A. (2019). Decision-making in schizophrenia: A predictive-coding perspective. *NeuroImage, 190*, 133–143. https://doi.org/10.1016/j.neuroimage.2018.05.074

Stringaris, A., et al. (2010). What's in a disruptive disorder? Temperamental antecedents of oppositional defiant disorder: Findings from the Avon longitudinal study. *Journal of the American Academy of Child & Adolescent Psychiatry, 49*(5), 474–483.

Sturm, W., & Willmes, K. (2001). On the functional neuroanatomy of intrinsic and phasic alertness. *NeuroImage, 14*, 76–84.

Stuss, D. (2006). Frontal lobes and attention: Processes and networks, fractionation and integration. *Journal of the International Neuropsychological Society, 12*(2), 261–271. https://doi.org/10.1017/S1355617706060358

Suarez, L., Markello, R., Betzel, R., & Misic, B. (2020). Linking structure and function in macroscale brain networks. *Trends Cognitive Science, 24*(4), 302–315. Epub 2020/03/12. pmid:32160567.

Suslow, T., Hußlack, A., Kersting, A., & Bodenschatz, C. M. (2020). Attentional biases to emotional information in clinical depression: A systematic and meta-analytic review of eye tracking findings. *Journal of Affective Disorders*, 632–642. https://doi.org/10.1016/j.jad.2020.05.140

Sylvester, C., Corbetta, M., Raichle, M., Rodebaugh, T., Schlaggar, B., Sheline, Y., et al. (2012). Functional network dysfunction in anxiety and anxiety disorders. *Trends in Neuroscience, 35*(9), 527–535. https://doi.org/10.1016/j.tins.2012.04.012

Tannock, R. (2000). Attention-deficit/hyperactivity disorder with anxiety disorders. In T. Brown (Ed.), *Attention-deficit disorders and comorbidities in children, adolescents, and adults* (pp. 125–170). American Psychiatric Publishing, Inc.

Tegelbeckers, J., Bunzeck, N., Duzel, E., Fletchtner, H., & Krauel, K. (2015). Altered salience processing in attention deficit hyperactivity disorder. *Human Brain Mapping, 36*(6), 2049–2060. https://doi.org/10.1002/hbm.22755

Thapar, A., Langley, K., O'Donovan, M., & Owen, M. (2006). Refining the attention deficit hyperactivity disorder phenotype for molecular genetic studies. *Molecular Psychiatry, 11*(8), 714–720.

Theeuwes, J. (2018). Visual selection: Usually fast and automatic; Seldom slow and volitional. *Journal of Cognition, 1*(1). https://doi.org/10.5334/joc.13

Thompson, R. (1994). Emotion regulation: A theme in search of definition. *Monographs of the Society for Research in Child Development, 59*(2/3), 25–52. https://doi.org/10.2307/1166137

Tognoli, E., & Kelso, J. (2013). The metastable brain. *Neuron, 81*. https://doi.org/10.1016/j.neuron.2013.12.022

Treisman, A. (1969). Strategies and models of selected attention. *Psychological Review, 76*, 282–299.

Treisman, A., & Gelade, G. (1980). A feature-integration theory of attention. *Cognitive Psychology, 12*, 97–136.

Tsuchida, A., & Fellows, L. (2013). Are core component processes of executive function dissociable within the frontal lobes? Evidence from humans with focal prefrontal damage. *Cortex, 49*(7), 1790–1800. https://doi.org/10.1016/j.cortex.2012.10.014

Tsujimoto, S. (2008). The prefrontal cortex: Functional neural development during early childhood. *The Neuroscientist, 14*, 345–358.

Tyukin, I., Gorban, A., Calvo, C., Makarova, J., & Makarov, V. (2018). High-dimensional brain: A tool for encoding and rapid learning of memories by single neurons. *Bulletin of Mathematical Biology, 81*(11), 4856–4888. https://doi.org/10.1007/s11538-018-0415-5

Ungerleider, L., & Haxby, J. (1994). What' and 'where' in the human brain. *Current Opinion in Neurobiology, 4*, 157–165.

van den Heuvel, M., & Sporns, O. (2013). Network hubs in the human brain. *Trends in Cognitive Sciences, 17*(12), 683–696. https://doi.org/10.1016/j.tics.2013.09.012

van den Heuvel, M., et al. (2018). Hubs in the human fetal brain network. *Developmntal Cognitive Neuroscience*. https://doi.org/10.1016/j.dcn.2018.02.001

van der Meer, J. M., Oerlemans, A. M., Van Steijn, D. J., Lappenschaar, M. G., De Sonneville, L. M., Buitelaar, J. K., & Rommelse, N. N. (2012). Are autism spectrum disorder and attention-deficit/hyperactivity disorder different manifestations of one overarching disorder? Cognitive and symptom evidence from a clinical and population-based sample. *Journal American Academy Child Adolescent Psychiatry, 51*(11), 1160–1172. https://doi.org/10.1016/j.jaac.2012.08.024

Vanunu, Y., Hotaling, J., Le Pelley, M., & Newell, B. (2021). How top-down and bottom-up attention modulate risky choice. *PNAS, 118*(39). https://doi.org/10.1073/pnas.2025646118

Visintin, E., De Panfilis, C., Antonucci, C., Capecci, C., Marchesi, C., & Sambataro, F. (2015). Parsing the intrinsic networks underlying attention: A resting state study. *Behavioural Brain Research, 278*, 315–322. https://doi.org/10.1016/j.bbr.2014.10.002

Vossel, S., Geng, J., & Fink, G. (2014). Dorsal and ventral attention systems: Distinct neural circuits but collaborative roles. *The Neuroscientist, 20*(2), 150–159. https://doi.org/10.1177/1073858413494269

Wang, B., & Theeuwes, J. (2020). Salience determines attentional orienting in visual selection. *Journal of Experimental Psychology: Human Perception and Performance, 46*(10), 1051–1057. https://doi.org/10.1037/xhp0000796

Wang, W., Shen, J., Cheng, M., & Shao, L. (2019). An iterative and cooperative top-down and bottom-up inference network for salient object detection. In *Proceedings of the IEEE/CVF conference on computer vision and pattern recognition* (pp. 5968–5977). CVF.

Wang, X., Zhou, W., & Zhu, X. (2020). Attention deficits in adults with major depressive disorder: A systematic review and meta-analysis. *Asian Journal of Psychiatry, 53*, 102359. https://doi.org/10.1016/j.ajp.2020.102359

Wang, Y., et al. (2024). The neural and genetic underpinnings of different developmental trajectories of attention-deficit/hyperactivity symptoms in children and adolescent. *BMC Medicine*. https://doi.org/10.1186/s12916-024-03449-1

Wasserman, L. W. (2024, April 19). Motor network and implications for neurodevelopmental disorders. Costa Mesa, California, USA.

Wasserman, T., & Wasserman, L. (2013). Toward an integrated model of executive functioning in children. *Journal of Applied Meuropsychology: Child, 2*(2), 88–96. https://doi.org/10.1080/21622965.2013.748394

Wasserman, T., & Wasserman, L. (2015). The misnomer of attention deficit hyperactivity disorder. *Applied Neuropsychology: Child*, 1–7. https://doi.org/10.1080/21622965.2015.1005487

Wasserman, T., & Wasserman, L. (2020). *Motivation, effort and the neural network model*. Springer Nature.

Weber-Goericke, F., & Muehlhan, M. (2019). A quantitative meta-analysis of fMRI studies investigating emotional processing in excessive worriers: Application of activation likelihood estimation analysis. *Journal of Affective Disorders, 243*, 348–359. https://doi.org/10.1016/j.jad.2018.09.04

Weiss, M. (2000). *Weiss functional impairment rating scale (WFIRS) self-report*. University of British Columbia.

Wells, A. (2009). *Metacognitive therapy for anxiety and depression*. Guilford Press.

Wells, A., & Fisher, P. (2015). *Treating depression: MCT, CBT, and third wave therapies*. John Wiley & Sons.

Wen, X., et al. (2019). First-year development of modules and hubs in infant brain functional networks. *NeuroImage, 185*, 222–235. https://doi.org/10.1016/j.neuroimage.2018.10.019

Willcut, E., Doyle, A., Nigg, J., Faraone, S., & Pennington, B. (2005). Validity of the executive function theory of attention-deficit/hyperactivity disorder: A meta-analytic review. *Biological Psychiatry, 57*(11), 1336–1346.

Wolfe, J., Cave, K., & Franzel, S. (1989). Guided search: An alternative to the feature integration model for visual search. *Journal Experimental Psychology: Human Perception and Performance, 15*(3), 419–433.

Woodman, G., Luck, S., & Schall, J. (2007). The role of working memory representations in the control of attention. *Cerebral Cortex, 17*, i118–i124. https://doi.org/10.1093/cercor/bhm065

Wu, Z., Wang, P., Liu, J., Liu, L., Cao, X., Sun, L., et al. (2023). The clinical, neuropsychological, and brain functional characteristics of the ADHD restrictive inattentive presentation. *Frontiers in Psychiatry, 14*, 1099882. https://doi.org/10.3389/fpsyt.2023.1099882

Xuan, B., Mackie, M., Spagna, A., Wu, T., Tian, Y., Hof, P., & Fan, J. (2016). The activation of interactive attentional networks. *NeuroImage, 129*, 308–319. https://doi.org/10.1016/j.neuroimage.2016.01.017

Yang, W., Ding, Z., Dai, T., Peng, F., & Zhang, J. (2015). Attention bias modification training in individuals with depressive symptoms: A randomized controlled trial. *Journal of Behavior Therapy and Experimental Psychiatry, 49*(part A), 101–111. https://doi.org/10.1016/j.jbtep.2014.08.005

Yao, Y., Xuan, Y., Wu, R., & Sand, B. (2020). Regulatory effects of reward anticipation and target on attention processing of emotional stimulation. *Frontiers on Psychology*. https://doi.org/10.3389/fpsyg.2020.01170

Yaroslavsky, I., Allard, E., & Snachez-Lopez, A. (2019). Can't look away: Attention control deficits predict rumination, depression symptoms and depressive affect in daily life. *Journal of Affective Disorders, 245*, 1061–1069. https://doi.org/10.1016/j.jad.2018.11.036

Yiend, J. (2010). The effects of emotion on attention: A review of attentional processing of emotional information. In J. de Houwer & D. Hermans (Eds.), *Cognition & Emotion. Current research and theories*. Psychology Press.

Zang, Y., Li, Q., Wang, Z., Liu, X., & Zheng, Y. (2017). Temporal dynamics of reward anticipation in the human brain. *Biological Psychology, 128*, 89–97. https://doi.org/10.1016/j.biopsycho.2017.07.011

Zhang, Y. C., Xin, Y., Peng, B., & Liu, S. (2023). Norepinephrine system at the interface of attention and reward. *Progress in Neuro-Psychopharmacology and Biological Psychiatry, 125*. https://doi.org/10.1016/j.pnpbp.2023.110751

Zhao, T., Mishra, V., et al. (2019). Structural network maturation of the preterm human brain. *NeuroImage, 185*, 699–710. https://doi.org/10.1016/j.neuroimage.2018.06.047

Zhao, Y., Yang, L., Gong, G., Cao, Q., & Liu, J. (2022). Identify aberrant white matter microstructure in ASD, ADHD and other neurodevelopmental disorders: A meta-analysis of diffusion tensor imaging studies. *Progress in Neuro-Psychopharmacology and Biological Psychiatry, 113*, 110477. https://doi.org/10.1016/j.pnpbp.2021.110477

Zohary, E., Shadlen, M., & Newsome, W. (1994). Correlated neuronal discharge rate and its implications for psychophysical performance. *Nature, 370*, 140–143.

Zorcec, T., & Jordanova, N. (2010). ADHD as an executive dysfunction. Contributions, section biological. *Medical Science, XXXI/2*, 171–181. 159.922.7:616.8–008.6.

Zuo, N., Yang, Z., Li, J., & Jiang, T. (2018). Core networks and their reconfiguration patterns across cognitive loads. *Human Brain Mapping, 39*(9), 3546–3557. https://doi.org/10.1002/hbm.24193

Index

T. Wasserman, L. D. Wasserman, *Disorders of Attention*, Neural Network Model: Applications and Implications, https://doi.org/10.1007/978-3-031-78732-4

V

W

The manufacturer's authorised representative in the EU is Springer
Nature Customer Service Centre GmbH, Európaplatz 3, 69115 Heidelberg,
Germany. If you have any concerns regarding our products, please
contact ProductSafety@springernature.com

Printed and bound by CPI Group (UK) Ltd, Croydon, CR0 4YY
01/10/2026
02225914-0006